# Ethics Hanbook
# for the Space Odyssey

Jacques Arnould

T0126332

# Ethics Hanbook
# for the Space Odyssey

### Jacques Arnould

### Translated from French by Yanette Shalter

Adelaide
2019

Cover design: Myf Cadwallader
Layout by Extel Solutions
Text: Minion Pro size 10 &11

ISBN:   978-1-925612-17-2    soft
        978-1-925612-18-9    hard
        978-1-925612-19-6    epub
        978-1-925612-20-2    pdf

Published by:

An imprint of the ATF Press Publishing
Group owned by ATF (Australia) Ltd.
PO Box 504
Hindmarsh, SA 5007
ABN 90 116 359 963
www.atfpress.com
Making a lasting impact

'Man must rise above the Earth—to the top of the atmosphere and beyond—for only thus will he fully understand the world in which he lives.' Socrates.

'Two things fill the mind with ever new and increasing admiration and awe, the more often and steadily we reflect upon them: the starry heavens above me and the moral law within me.' Immanuel Kant, *Critique of Practical Reason*

# Table of Contents

# Introduction:
# Space and Ethics Cannot be Dissociated

'Why should ethics be of particular concern to space agencies and industries? Of course, ethical behavior should be of major concern to all of us in all occupations, but because of the special nature of space activities—their expense, their danger, their public visibility, and their unique challenges and opportunities—space and ethical reflection must go together.'[1]

We must take seriously what Jim Dator says about the necessity of associating space and ethics. When he and I met in Valparaiso in 2000, during the summer session at the International Space University (ISU), he was advising his students to think about the ethical questions potentially raised by space exploration. I remember his questioning them about the opportuneness of informing the astronauts of a space shuttle in the event a major disaster were to threaten them, one that only ground control would know about. The Columbia tragedy occurred two and a half years later. Professor Dator eventually drew me into this teaching. In various forms, during summer sessions or Master's degree courses at ISU, I offered to sensitise students to the ethical stakes of space activities. I have also had the opportunity to speak about this subject at other academic institutions (universities, engineering schools), and I continue doing so in France and abroad.

My experience is different from that of Jim Dator. This teacher and researcher at the University of Hawaii has developed an in-depth understanding of relationships between space and human societies, based on his expertise that focuses on our societies' social and politi-

---

1. James A Dator, *Social Foundations of Human Space Exploration* (New York: Springer-ISU, 2012), 79.

cal future, on utopias, and even on the role of the media. As for me, I have acquired a more practical proficiency and experience at the Centre National d'Etudes Spatiales (CNES), the French space agency, where, since 2001, I have held the position of ethics expert. This is the experience I share when I address students and other audiences.

I strive to show how, far from being useless or even disturbing for space activities, ethics can actually provide support in various ways. They can open the door to a process of collective intelligence among the various communities of space actors, or they can serve the purpose of informing the citizens of our countries, who are still today the main contributors to space activities (even taking into account the emergence of NewSpace). I also seek to show how space offers humankind singular opportunities to question itself, its inspirations, expectations, and fears. Thanks to space, our humanity continues to grow. And this deserves our full attention.

# Chapter 1
# Multiple Spaces

Gunpowder is probably one of China's most famous inventions, together with the seismograph, the compass, and printing. It was used for the first time in 919 as a detonating agent and in 970 as a propellant. In the sixteenth century, when the Mandarin Wang Hou first imagined using it to become airborne, he disposed of no satisfactory technical solution to simultaneously ignite the forty-seven rockets of his flying machine, which was also equipped with two kites. He thus resorted to forty-seven slaves, each of whom was charged with igniting one rocket. One of them inadvertently set Wang Hou's flying machine on fire. The Mandarin died in flames and, instead of effectively rising to the sky, became part of the legend of space exploration. When Chinese engineers and dignitaries proceeded to launching the first Chinese habitable spaceship from the Gobi desert, in November 1999, they knew the story of their distant precursor. They also knew that, in the space race, they had been preceded by other nations: the first modern-day 'sons of the sky' were not descendants of the emperors of China, but were all born in the West, either Russian or American.

We must take stock of this simple historical observation. We cannot question the essential purposes, conditions, and motivations of our ventures, in terms of space exploration and usage, without taking into account the diversity of ways of perceiving, comprehending, and representing the sky, conveyed to us by present and past cultures. There is not only one space, but many.

## Sacred skies

One thing is certain: up to this day, not a single human civilization has ignored the starry firmament. Archeoastronomy has found proof of this among the first human remains, on the walls of prehistoric caves or on carved, engraved objects. The drawings of the Moon and the Pleiades in the Caves of Lascaux, France, are more than 15,000 years old. The Nebra sky disk made of bronze and gold, discovered in German Saxony at the end of the twentieth century, is believed to date back to 1600 BC: it represents the Moon, the Sun, and most likely the Pleiades.

Should we search even further back for the roots of such a deep interest, namely in the biological or animal breeding grounds from which our species has evolved? The wolf's nocturnal howling has given rise to many legends, although this behavior can easily be explained. When the animal lifts its head to howl, its vocalisation becomes louder and carries farther. The Moon very likely has nothing to do with it. Must one rule out, however, possible effects of luminous and celestial variations on more 'simple' creatures, provided they are endowed with some sort of photosensitive captor? How can we be certain that some of our primate cousins don't raise their eyes to the sky to enjoy its cold yet bright beauty, to let their animal consciences become imbued with the presence of an elsewhere, of another world, other beings, either similar to themselves or completely different? We shall leave this matter to biologists and ethologists of all sorts and simply state here that the sky has always fascinated human beings.

I use the word 'fascination' to designate the double motion of attraction/repulsion, the simultaneous feeling of happiness and anguish, of peace and fear, that the human spirit can experience at the sight of the starry sky. This is how the Egyptians probably felt when they depicted their goddess Nout, her body spangled with stars, arched above the Earth, her laughter triggering thunder and her tears turning to rain. Every evening, the goddess swallows the Sun to give birth to it again every morning. Thus, Nout is the protector of the living and the dead: the Egyptian tradition compares her to the sow who devours her own offspring. Can we, today, claim to no longer have such fascination simply because we now have the means to observe the most distant corners of our universe, or because humanity has left footprints on the Moon? The mere announcement of a solar eclipse, or information *a posteriori* of an asteroid passing by close to the Earth, can awaken the most ancestral hopes and fears in our scientific, rational, or religious minds. Hence, Immanuel Kant was right to evoke the veneration that flows from the sight of the night sky: to human perception, the sky possesses all the attributes of the sacred.

## Forbidden cosmos

Sacredness comes together with notions of sacralization and prohibition and, in contrast, of transgression and profanation. This also applies to the sky. Nonetheless, it was probably Aristotle, even more than religion, who set the most solid, impassible boundaries between heaven and earth (I will leave hell aside). That the Earth should occupy the center of reality flows from the most common of experiences: the Sun 'rises' and 'sets', passing above our heads; the stars follow paths and our Earth appears to be at the center of these. In the absence of other data or constraints, how could one possibly envisage and defend any idea other than geocentrism? This theory of Aristotle's, which dominated Western thought and culture for a long time, is associated with a strict dualism.

If the Earth is the centre of the world, if it is its *omphalos*, its navel, that doesn't mean it is a summit, a place that confers nobility and prestige to whoever stands on it. On the contrary, it is rather a cesspool, a place where all impurities are gathered, including the human species, which only strives in mire, chaos, and war. If the Sun turns on its crystal sphere like a gigantic projector meant to light up the earthly stage, it is for the deity, all-mighty Creator or Great Architect, not to lose a single crumb of the human comedy. Unless these spheres, which also carry the planets, are also in charge of hiding from any celestial gaze the filth and vice accumulated on Earth. There is no point in trying to exclude either of these two interpretations of the ancient organization of the world, with its opposition between the cosmos (the Greek term designates an 'orderly, harmonious system') and the Earth as it was first elaborated by Aristotle, then defended by philosophical and theological traditions since the dawn of modern times. Although they seem opposite to each other, they coexisted and survived in Western cultures, following human kind's whims and traditions up to the seventeenth century.

The consequence of this understanding, or rather of this cosmic interpretation of the sky, was that humans were denied access to it.

There was no question for them to go join the celestial spheres, or at least not with their bodies or during their earthly lives. Only the wisest, the most saintly or spiritual ones among them could hope to go to those places, after having led perfect, ascetic, nearly angelical lives. This ban was so efficient that it even impacted Western imagination. Indeed, literary works relating a human being's (bodily) voyage through the sky are so rare that they are the exceptions that prove the rule.

Elsewhere, in other cultures, even in places where Aristotle had no influence, dreams of reaching the Moon don't seem to have filled people's nights and imaginations. There is often mention of celestial, divine beings who populated the skies and stars, or who visited our world, but there are no stories of humans embarked on celestial expeditions. The sky was not always a forbidden place, but it remained inaccessible.

---

### The celestial ladder

As an allegory of spiritual ascension, the celestial ladder appeared in Christian art as early as the fourth century AD. It was developed throughout the Middle Ages and constantly renewed. If it expresses a fervent longing for the sky, it also conveys fundamental elements of medieval religious thought. The world was created by God according to a dualistic process of separation between heaven and earth; the moral fall of humanity (called 'sin') is mirrored by Christ's salvation when he 'came down from Heaven' to give believers the chance to rise again. This journey back upwards is a gradual spiritual progression.

The lesson of the celestial ladder is a paradoxical one: the path towards elevation is also a humbling one; rising through contemplation comes with descending through compassion and charity. And to this, we may add the idea that sanctity is not a state granted at birth, but the consequence of a life-long personal quest . . . or exploration!

---

### Space of men

In 1543, Nicolaus Copernicus died. As the story goes, he was on his deathbed when he received the book in which he defended his theory of heliocentrism. He placed the Sun, and not the Earth, at the center of the universe. In fact, even before the Polish ecclesiastic, other thinkers had elaborated and taught heliocentric cosmologies: Philolaos in the fifth century BC, Heraclides Ponticus in the middle of the fourth, and Aristarchus of Samos shortly after him. All defended a heliocentric vision of the world—in vain. Indian and Muslim astronomies also convey heliocentrism. In the medieval West, after Jean Buridan and Nicole Oresme, Cardinal Nicholas of Cusa also questioned the prevailing theory of geocentrism. He wrote 'Why should we hesitate any longer to grant it [the Earth] a mobility which, by nature, matches its form, rather than try to shake the entire universe of which we do not, and cannot, fathom the boundaries?' In his work *De docta ignorantia*, published in 1440, he went so far as to claim 'Life exists here on Earth in the shape of humans, animals, and plants, but let us suppose it exists, in a more elevated form, in the solar and stellar regions'[2]. Sixty years later, Leonardo da Vinci formulated the hypothesis that the Earth might be a celestial body of the same nature as the Moon. And Giordano Bruno, before he died at the stake in Rome in 1600, spoke of infinite numbers of worlds and earths. Yet in spite of this, when, in 1610, Galileo published his treatise *Sidereus Nuncius, The Sidereal Messenger*, exposing the results of the observations he had made that same winter in Padua with a telescope of his own making, he really came across as a revolutionary.

His observations, namely that the Moon was not smooth and polished, but just as rough and uneven as the Earth, and that Jupiter was accompanied by three, then four satellites, led him to understand that both terrestrial and celestial worlds were made of the same matter and governed by the same laws. The world was not separated into earth and cosmos. The world was one, it was a universe. From there on, nothing could any longer justify that the Earth, our Earth, should occupy the center of it. Hence, it became a planet among others, turning with them, and just like them, around the Sun.

In Prague, where he held the position of imperial mathematician at the court of Rudolf II, Johannes Kepler confirmed Galileo's observations and conclusions. The ancient crystal spheres were shattered, the world became a boundless universe, devoid of a centre and even

---

2. *Cf* Duhem, Pierre. *Le système du monde. Histoire des doctrines cosmologiques de Platon à Copernic*, t. X (Paris: Hermann, 1965), 324.

of a circumference as it now seemed endless. There was no longer
any position, any place that gave him who stood in it a sense of pre-
eminence or decline. Humanity was left to digest its humiliation or to
rejoice in its rehabilitation. In a letter he wrote to support *The Sidereal
Messenger* and Galileo's ideas, Kepler went further than merely defending his Italian colleague's theories. In fact, he put in writing the very
conception of astronautics (or in other words, of space travel), for he
was convinced that 'There will certainly be no lack of human pioneers
when we have mastered the art of flight'. The century that ended a
decade earlier had been a century of discovery, exploration, and early
colonization of the Americas. Ninety years earlier, humanity had for
the first time successfully gone around the world. Thus, Kepler had
no doubt that the perspective of another New World, a celestial one
this time, would soon prompt a similar burst of enthusiasm among
the boldest of men. It became obvious for the astronomer to compare
it with the mastery of navigation and the feats of European sailors:
'Who would have guessed that navigation across the vast ocean is
less dangerous and quieter than in the narrow, threatening gulfs of
the Adriatic, or the Baltic, or the British straits?' And Kepler carried
on 'Let us create vessels and sails appropriate for the heavenly ether,
and there will be plenty of people unafraid of the empty wastes'. In
spite of this, he still seemed to harbor some doubts regarding the time
required to build the first spaceships. Consequently, full of common
sense, he added 'In the meantime, we shall prepare, for the brave sky-
travellers, maps of the celestial bodies – I shall do it for the Moon and
you, Galileo, for Jupiter.'[3] Resting resolutely on the observations of his
colleague in Padua and on their common conclusions, and enthused
by the perspective that humankind might some day escape its earthly
prison, Kepler was convinced that, henceforth, nothing would be too
high nor too far for humanity to decide to undertake the voyage. The
astronomer entrusted engineers with the task of inventing navigation towards the stars, and he dedicated himself to the elaboration
of maps which could be of use to the first cosmic explorers and navigators. This cartographic endeavor seemed indispensable to him in
order to foresee and reveal, even from afar, the worlds and islands,
the reefs and pitfalls that the space *conquistadores* might encounter

---

3.   Quoted in Arthur Koestler, *The Sleepwalkers: A History of Man's Changing Vision
     of the Universe* (London: Pelican, 1968), 378.

throughout their navigation. These maps were even meant to arouse interest, to feed the audacity of those who would go explore distant worlds still to be discovered. Kepler, who practiced astrology to earn a living, was convinced of this. The time would soon be over when humans, punished by some celestial power to remain imprisoned on Earth, felt content with reading their futures in the stars. He claimed enthusiastically that tomorrow, humankind would go spell its destiny amidst the stars! The sky would no longer be the sole realm of gods and angels. It could henceforth also become the space of men.

## From dreams to reality

The revolution triggered by Galileo and Kepler did not only shake the fields of science and foreshadow the astronautic techniques that would appear three and a half centuries later, it also freed people's imaginations: voyages to the Moon or the stars became a topic in many works of Western literature and culture. Even before he wrote his reply to Galileo, his *Conversation with the Sidereal Messenger*, Kepler himself wrote a work of fiction entitled *Somnium, seu opus posthumum de astronomia*, also called *The Dream, or Lunar Astronomy*. There, he evoked a trip to a mysterious island, Levania, which represented the Moon. However, it dealt more with witchcraft and the power of spirits and demonic influences than with astronomy and astronautics . . .

Francis Godwin is often considered to be the first modern author to have imagined a voyage to the Moon. In 1638, in *The Man in the Moon*, he shared his vision of an enchanting lunar nature and a more accomplished humanity than our own. Twenty years later, Cyrano de Bergerac presented *Les États et Empires de la Lune* (1657) followed by *Les États et Empires du Soleil* (1662) (translated into English in 1687 under the title *The Comical History of the States and Empires of the Worlds of the Sun and Moon*). In 1765, Marie Anne de Roumier published the seven volumes of *Voyage de Milord Céton dans les Sept Planètes (Lord Seton's Voyage Among the Seven Planets)*, that recount a space saga. In 1835, it was Edgar Allan Poe's turn to send Hans Pfaall to the Moon aboard a revolutionary balloon, in *The Unparalleled Adventure of One Hans Pfaall*.

The systematic exploration of the globe, which had started at the end of the 15th century, ended in the nineteenth century. Henceforth, human beings would be left with the sky to quench their curiosity and their thirst for exploration. Space travel went from being a mere literary exercise to a scientific project. The telescope, first used in the early seventeenth century, was promoted to the rank of vehicle: through its lenses (and with a good deal of imagination) astronomers could scrutinise the Moon, planets, and stars. Following the footsteps laid in Kepler's *Conversation*, they took on the project of establishing a cartography of the surface of Mars, as precisely as in maps of the Earth . . . or at least, that was their aim. Angelo Secchi, Giovanni Schiaparelli, and Percival Lowell even drew extraordinary networks of canals on the surface of the red planet: science is not impervious to imagination . . . at its peril. The nineteenth century was a fascinating period! It greeted with similar enthusiasm Jules Verne's *Voyages extraordinaires* (*Extraordinary Voyages* or *Extraordinary Journeys*) and Camille Flammarion's publications. In his *Pluralité des mondes habités* (*The Plurality of Inhabited Worlds*), the French astronomer and populariser claimed that 'the Earth disposes of no obvious pre-eminence in the solar system, indicating that it should be the only inhabited world; astronomically speaking, the other worlds are just as likely to harbor life'. As Kepler's worthy successor, he associated astronomy and astronautical invitation. But real space travel still remained a matter of fiction.

Needless to say, the twentieth century suffered no shortage of imagination or authors to express it. From Herbert G Wells (*The War of the Worlds* in 1898) to Arthur C Clarke (*2001, a Space Odyssey* in 1968), the literary vein of space fantasy was broadly exploited. Wells is even considered the inventor of modern science-fiction. Thanks to him, 'scientific marvels' such as the time machine, the matter transmuter, or hyperspace entered the public domain. Clarke, who was a member of the British Interplanetary Society, published in 1939 an article entitled '*We* Can Rocket To The *Moon*—Now!'. He went on to elaborate the concept of geostationary satellites before becoming one of the most renowned science-fiction authors of the second half of the 20th twentieth century. Many fields of cultural and artistic creativity turned towards space. Film-making started enriching the resources already offered by literature and plastic arts, as cinema was inaugurated with great ceremony in 1902 by George Méliès, with his film *Voyage dans la Lune* (*A Trip to the Moon*). Up to the middle of the twentieth century, the sky, which had progressively lost its religious connotations in the West, remained the setting and/or the motif for major works of imagination.

On October 4, 1957, a key event in the history of space adventure occurred. After that, it would no longer be only a matter of imagination or astronomical research. With the launching of the first *Sputnik*, space adventure finally entered the age of technical achievements, of effective space conquest. This event had a far-reaching impact that 'nothing, not even the fission of the atom, would be able to eclipse', as philosopher Hannah Arendt wrote in her book *The Human Condition*,[4] before she quoted a few words by one of the founders of modern astronautics, Konstantin Tsiolkovsky: 'Earth is the cradle of humanity, but one cannot live in a cradle forever.'[5] Cosmic travel, which Western tradition had imagined since 1610, had become if not an immediate reality, at least a conceivable one. Human beings could now legitimately claim to be citizens of the cosmos, children of the stars, 'Spatiopithecus'. Nonetheless, the realm of space remained bound to the imaginary world: in his book *Space and the American Imagination*, Howard McCurdy showed how writers and promotors of the American space program relied on prominent cultural figures such as Walt Disney to arouse interest and gain supportive public opinion[6]. Far from being a North-American exception, this connection between dream and reality regarding space matters exists in other countries as well. To give only one example: in Japan, the common interest for lunar missions can easily be explained by the cultural importance of the story of Kaguya, the 'radiant bamboo princess'. 'Fallen' from the Moon and taken in by a poor bamboo cutter, she strived to escape her suitors, all enthralled by her beauty, in order to return to the planet of her birth . . .

### Other universes?

Today, space still requires the resources of our imagination, not only to support scientific research and technological innovation, but also to feed ethical reflection. Let me reiterate that the latter may not be limited to judging the past or examining present human actions. Ethical committees are not to replace the tribunal of history. On the

---

4. *Cf* Annah Harendt, *Condition de l'homme moderne* (Paris: Calmann-Lévy, 1983), 33–34.
5. In a letter to the engineer, Boris Vorobiev, on August 12, 1911.
6. *Cf* Howard E McCurdy, *Space and the American Imagination* (Washington/London: Smithsonian Institution Press, 1997).

contrary, they should rather turn towards the future and, therefore, they depend upon imagination. The behaviors of tomorrow's human beings are already in the process of being created today.

It might seem unusual to associate ethics with a prospective dimension. A few years ago, the president of the Comité consultatif national d'éthique français (CCNE) (*French National Ethics Consulting Committee*) was surprised when I asked him how much importance he gave imagination in the context of his committee's work, whereas I consider it to be indispensable, especially in space matters. What we call 'space' is above all the product of our actions, meaning the result of what we first imagined we could do. This is why I spoke of the 'space of men'. Nonetheless, I know there are still many *terrae incognitae*, unknown territories, worlds of which we know nothing yet, have not an inkling of an idea . . . in short, other universes.

There is no question of speculating about the morals and ethics adopted by hypothetical extraterrestrial beings potentially gifted with intelligence, conscience, or reason, nor even to discuss the rights they would be granted by our institutions. Such an endeavor would be useless insofar as we haven't got the slightest proof of their existence. That being said, we can wonder about the duties that should be ours vis-à-vis these hypothetical creatures and their extraterrestrial environments, and also about what we would owe ourselves in the event of a close encounter of the third kind. This is a prospective job, a task that requires interrogating and analyzing the reflections of our own ideals, our own fears, when confronted with the possible existence of other living, conscious, and intelligent—nonhuman—beings. What can we learn about the space of others?

Space is not singular. On the contrary, there are multiple spaces, depending on circumstances, periods in time, cultural contexts. Determining, defining which one we are thinking of or referring to is probably the key prerequisite when questioning the purposes of our astronautic activities, their conditions, and their consequences. In short, whenever we wish to adopt an ethical stance.

# Chapter 2
# What Are Ethics?

---

Ethics have been trendy for over thirty years. This familiar subject among ancient philosophers, a fundamental building-block of philosophical works and religious teachings, remains to this day frequently bound to the context of academic seminaries or to the institutionalised framework of ethical committees. Nowadays, however, it also exists outside these habitual fields for it is claimed by experts and circles that had previously been alien to it. Henceforth, one can hear of ethics in management, journalism, commerce, finance, and even fashion. In fact, the term 'ethical' is sometimes used as a social foil or a sales argument. At first sight, ethics seem to have become indispensable for the conduct of human affairs. If we look into the etymology of the term 'ethics', this fact shouldn't surprise us, and we should even welcome it. The Latin word *mors* and the Greek word *ethos* both denote the same thing: they refer to a manner of behaving, to the mores specific to each human individual and society. In other words, nothing involving human activities should escape undergoing reflection, that is to say, an ethical process.

---

### An academic typology of ethics

Ethics are

1.  the philosophical study of the moral value of human conduct and of the rules and principles that ought to govern it (*meta-ethics*);
2.  a social, religious, or civil code of behavior considered correct, especially that of a particular group, profession, or individual (*normative ethics*);
3.  the moral fitness of a decision, course of action, etc (*applied ethics*).

---

### Extraterrestrial ethics?

Let's beware that this apparent success of ethics might be merely due to the diversity and grey areas of their current definitions. In fact, other voices are being raised, though more discretely, regarding ethics' harmful effects on scientific and technical progress. Here is one example. In January 1999, when Albert Ducrocq was informed of the ethical initiative of the European Space Agency (ESA)—which I will come back to—he published a very critical article in the weekly magazine *Air & Cosmos*. This famous French scientific journalist defended the idea that spatial debris represents a near-zero danger for Earth; as for the speeches about the samples of Martian soil brought back to Earth (potentially containing living organisms), he added that they served no other purpose than to terrorise the public and hide the true scientific interest of such a mission. He concluded that undertaking an ethical process, elaborating a sort of space charter 'risks giving credit to the theory according to which astronautics not only constitute a useless expense, but also create problems capable of endangering our planet'. In other words, the ESA's good intention 'could very well do a disservice to the space program'.[1] Ducrocq's stance has the merit of being clear: there are plenty of arguments to defend the idea that space should be spared most of the ethical questioning with which other technological fields are presently confronted (I am namely thinking of biology and medicine, or nuclear activities, that are subject to ethical reflections, public debates, political and judicial decisions, etc).

What should we make of this 'anti-ethical' position regarding space? Does it suffice that space activities are extraterrestrial for them to claim an extraterritoriality or a specificity in terms of ethics? In short, should ethical matters be broached in a specific way when it comes to space? I do not believe this stance to be reasonably defendable, especially if its aim is to exempt astronautics from the ethical imperative which prevails in this day and age (even though it isn't common in all societies, in all the nations of the world). I rather believe that these criticisms and expressions of mistrust, which cannot be overlooked, stemmed from a mistaken approach, a partial understanding of what an ethical process really is.

---

1.  Albert Ducrocq, 'Éthique spatiale. Une bonne intention qui pourrait fort mal servir la cause de l'espace (Space ethics. A good intention that could very well do a disservice to the space program)'. *Air & Cosmos*. 1686 (1999): 39.

The presentation I am offering hereafter does not ignore the numerous papers and books that deal with the subject of ethics, morals, and their applications in the various fields of human activity. However, it relies first and foremost on my personal experience at the CNES.

## Ethics or questioning evidence

More than sixty years after it started, the exploration of space still appears as one of the most singular achievements of human history, and the astronauts' first steps on the surface of the Moon remain one of the most remarkable events of the twentieth century. This being said, many thinkers and actors of the spatial field deem that this is merely the continuity of a movement born several centuries earlier, a movement to which the astronomical revolution of the seventeenth century provided a crucial characteristic: its feasibility. Space suddenly appeared as an extra chapter to add to the odyssey of humanity, who never ceased exploring, but also conquering and colonizing new territories on the surface of the planet, before turning towards the air and, at last, towards the cosmos. When you ask astronauts why they are willing to risk their lives onboard a spaceship or on the surface of a planet other than Earth, they sometimes give the answer George Mallory used to give to those who wondered why he wanted to climb Mount Everest. 'Because it is there' would invariably be the reply of the British mountaineer, who was last seen alive on June 8 1924, on the Northern ridge of Everest, and whose body was found seventy-five years later, on May 1 1999, at 8,300 meters above sea level. As if traveling to space were also something completely self-evident.

There is no doubt that human existence is founded on a number of self-evident things. Exploring space or climbing a mountain seem secondary to those that are common to all human beings (and even to most living beings): eating, finding shelter, resting, healing one another, procreating, working, having fun, etc. These are imperatives to the extent that we can speak of determinism in their regard: we do not have the choice to perform them or to refuse them, but we can choose under what conditions we will engage in them. Our diets, dwellings, working and resting places, love practices and conjugal rites all bear the mark of our traditions and our culture, in short of our human singularity, compared to all other living beings which are subject to the same determinisms. We need not wonder why, but simply how we intend to realise them and what consequences they may have (for example, on our health or environment). Asking these questions already reflects a true sense of our responsibility, but it is still only a part of the ethical process. This process must also lead us to question the why of our actions, meaning their goal, their finality, their raison d'être. And this dimension is essential: ethics must, above all, be a way of questioning evidence.

Let's take the example of play. This activity, common to humans and animals, is sometimes qualified by thinkers as 'acting without a why'. It has no other end than itself. Playing is self-evident, it occupies a key position in the education of younglings and in the social life of many living beings. It merely requires setting rules or codes of conduct and abiding by them. The question of the finality of playing arises when it is associated with mercantile or bellicose intentions where it then becomes necessary to decide that 'the game's over!'

Let's take another example: health. The art of medicine and, more generally, the care bestowed upon oneself and others is rightfully considered as self-evident. Medical practices, whether common or experimental, can be submitted to an ethical process, to the evaluation by experts or ethical committees. The latter must approve them, suggest improvements or, in some cases, condemn and ban them. It is quite rare though that the finality of medicine should be questioned, because the value of health is obvious, as are the efforts people are willing to make to maintain or recover it. But actually, it might not be as easy to answer this question as first meets the eye. According to the constitution of the World Health Organization (WHO), health is 'a state of complete physical, mental, and social well-being, and it does not only consist in the absence of disease or infirmity'. The document specifies that health represents 'one of human beings' fundamental rights, regardless of their religion, political opinions, social or economic situation'. Though this definition is pertinent, it is incomplete. The practical understanding of health, as a patient or a care-giver might see it, must necessarily include some individual and subjective aspects. That is why health-related activities raise the most sensitive and least self-evident ethical issues. A few years ago, I was invited to speak on the occasion of a meeting between representatives of ethical committees working on the revision of French bioethical laws. Totally lacking competence in the medical field, my contribution was to remind everyone that, within the context of manned flights, the finality of spatial medicine consists of protecting the health of individuals whose physical condition is excellent, in spite of the conditions to which they have voluntarily submitted. Considering medical practice from this unusual perspective (which, in fact, is opposite to therapeutics) is in itself a way of questioning medical evidence.

Presented as an invitation to question evidence, ethics can thus not be reduced to a line-drawing exercise and, even less so, to a pretext for prohibition. Ethics go much further as they compel the people involved to focus on their essence, their motivations, their reasons for being, the conditions of their choices, and their ways of applying decisions. If we could agree on such a definition of ethics, we would no longer be saying or hearing things such as 'Is this attitude or action ethical?', but rather 'Have we broached this problem, or undertaken this action by questioning its reasons and conditions, by adopting an ethical stance?' In virtue of the ethical approach, we must consider the soundness and consequences of our actions before we undertake them. Michel Foucault advocates nothing else when he elaborates his own definition of ethics as 'a way of thinking and feeling, but also of acting and behaving, that simultaneously marks a belonging and appears as a task. Somewhat similar, perhaps, to what the Greeks used to call *êthos*'[2].

## Ethics or managing possibilities

The purpose of ethics is not to open up the 'space of possibilities', nor is their function to explore this space; science and technology, as well as philosophy and art take care of that. The conquest of the skies is a good illustration of this. During centuries, for the philosophical reasons I mentioned earlier, such an endeavor was deemed impossible by Western cultures. It was the modern astronomical revolution that first made space travel fathomable for humans, then engineers made it possible, and eventually, in the middle of the twentieth century, they actually achieved it. The history of humankind is punctuated by moments when humans, facing the risks with audacity, dared to envisage new courses of life or action and tried to realise them.

---

2. Michael Foucault, 'Qu'est-ce que les Lumières?', (1984) in *Dits et écrits. 1954–1988*, volume IV: 1980–1988 (Paris: Gallimard, 1994), 568.

Throughout time, there has been no lack of men and women such as Pierre-George Latécoère who, just before creating one of the first airlines in December 1918, explained to one of his pilots 'I've made all necessary calculations. They confirm what the specialists say: our idea cannot be realised. We only have one thing left to do: realise it!' I am convinced that the same spirit pervaded John F Kennedy's speeches on space ('We choose to go to the Moon in this decade and do the other things, not because they are easy, but because they are hard') and that today, it still drives the actors of NewSpace ('Failure is not an option', asserts Elon Musk). Ethics, as a questioning approach, do not precede the engineer's or the entrepreneur's claim 'We can do it!'. On the other hand, ethics must imperatively accompany the decision to realise or not to realise that which is realizable—or even what might not seem to be.

We're broaching a thorny subject here: the responsibility of scientists and engineers, especially towards the societies they belong to. This leads us to raise the question of neutrality in scientific processes and technological innovations. Players in these spheres often claim and defend neutrality with the following argument: 'It falls to us, researchers and engineers, to discover and innovate; it falls to societies and their leaders to define how our work can best be put to profit and, conversely, how its applications should be limited or even forbidden!' What should we make of such a claim to neutrality and non-responsibility?

It is not hard to imagine that researchers and engineers adopt this defensive stance in order to avoid accusations that could stem from some misuse, incident, accident, or catastrophe. Nowadays, people sometimes show an exaggerated tendency to resort to judicial treatment when they run into hazards of life, at both the individual and collective levels. It is perfectly understandable to try to take advantage of this trend, as it is natural to try to escape from it, to protect oneself from it: such reactions are part of what can be referred to as 'managing possibilities'. However, I do not believe this is a proper way to interpret the meaning of ethics when applied to a profession or a trade (a branch commonly called 'professional ethics'). The scientist's or engineer's responsibility is engaged, including towards his society (the organization for which he works, his country, or even humanity as a whole), at all stages of his work: choosing a field of research or innovation, formulating hypotheses, validating a technological process, etc. Let's remember that Charles Darwin postponed the publication of his findings regarding the origin of living species out of concern for the philosophical and religious consequences those findings might have. He never claimed his work was neutral. Thus, to oppose the argument of neutrality, we can propose three reasons defending the argument that researchers and engineers should assume a social responsibility.

The first reason rests on the need for expertise. The conduct of our actions, as individuals or as societies, requires three steps, three successive stages: knowledge, expertise, and decision-making. Accordingly, the scientist offers a vision and an explanation of the situation (which, in Greek, is called *theoria*), the expert describes the possible outcomes (a counter expertise can contradict his conclusions), and the decider, the political person in charge, decides what action to take in view of these possible outcomes. If one of these tasks is missing or neglected (for example, if the decider merely follows the expert's advice without adding his own concern for the common good), the worst can happen. So, neutrality is not an option: there must be responsibility. Each of the three aforementioned stages belongs to the process and is responsible for its (successful) completion.

The second reason is called the precautionary principle. A definition of this principle was introduced in the Rio Declaration on Environment and Development (June 1992): 'In order to protect the environment, the precautionary approach shall be widely applied by States according to their capabilities. Where there are threats of serious or irreversible damage, lack of full scientific certainty shall not be used as a reason for postponing cost-effective measures to prevent environmental degradation.' (Principle 15). As Jean-Jacques Salomon points out, 'The precautionary principle is essential in a world where science wants to make us believe that everything that is possible is desirable, merely because it is realizable.'[3] In other words, the precautionary principle lies at the foundation of ethics in the sense of managing possibilities, i.e. ethics that are capable of distinguishing what is possible from what is desirable and/or realizable. It combines the need for early threat detection (which comes under the responsibility of scientists and experts) and the undertaking of preventive action

---

3. Jean-Jacques Salomon, *Survivre à la science. Une certaine idée du futur* (Paris: Albin Michel, 1999), 303.

without waiting for scientific certainties to be established (which is the deciders' role). The precautionary approach can thus be summarised as follows: 'When in doubt, do not refrain from acting, but act as if you had been proven right.'

The third reason is a just understanding of the imperative of responsibility, which, as mentioned previously, cannot be reduced exclusively to its judicial dimension. The philosopher Hans Jonas noted that, until recently, humanity's knowledge and power—in other words, its 'space of possibilities'—were too limited to stretch the ethical questioning beyond the moral quality of actions with consequences limited in space and time. More specifically, there was no need to include our planet Earth in the awareness of a collective causality and responsibility, let alone a personal one. Times have changed, however. Now, humanity has the ability to impact great ecological balances and threaten the existence of species or entire ecosystems. Activities are no longer restricted to local, immediate, or present spheres; they affect global, distant, and future conditions, sometimes with irreversible consequences. How to react, wonders Jonas? Certainly not by holding the ideal of progress which would have us believe that all difficulties can be solved by science and technology. Jonas does not condemn these fields, but, according to him, hummankind must define and establish its own limits to its development and ambitions. Humanity must self-limit in order to preserve its integrity and that of the whole world. The growing risks of planet devastation and self-destruction of the human species require us to extend the scope of the categorical imperatives cited by Immanuel Kant to include humankind's relations with its environment and with the future of its species. Jonas warns us to 'act in such a way that the effects of your action are compatible with the permanence of genuine human life on Earth [. . .] Act in such a way that the effects of your action are not destructive for the future possibility of such life.'[4] Regardless of the limits of Jonas' definition of the principle of responsibility (in particular the need to define the very notion of life), it presents the advantage of introducing into every ethical process a global approach regarding the (earthly) environment and a concern for the future generations. These two perspectives must be taken seri-

---

4. Hans Jonas, *Le Principe Responsabilité. Une éthique pour la civilisation technologique* (Paris: Flammarion, 1990), 40.

ously in space ventures. Not to do so would widen the gap between the representation of the world constructed by space-related sciences and technologies on one hand, and the reality directly experienced by the average human on the other.

## Ethics or 'the aesthetics of the soul'

For over sixty years, many space programs have provided us with a singular and totally novel experience of nature and especially of our home-planet. The first spatial decade and, more specifically, the Apollo program were contemporary to the rise of ecological and environmental movements. That planetary awareness was not devoid of a certain aesthetic dimension.

Apparently, an appreciation of beauty is common to all human beings and, simultaneously, unique to each one of them—or at least specific to each culture. According to Kant's teachings, the paradox of Beauty lies in the fact that it is at the same time the most universal and the most unique thing, probably because it stems from a complex intertwining of reason and imagination. Therefore, it is not surprising to bring together ethics and aesthetics, as the French poet Pierre Reverdy did. He wrote 'Ethics are the aesthetics of the soul'. That is a way of saying that ethics cannot be reduced solely to principles developed by reason and understanding, even taken collectively. Ethics maintain an element of mystery, of inner experience. This is what Kant suggested in the excerpt of *Critique of practical reason* quoted as an epigraph to this essay: 'Two things fill the mind with ever new and increasing admiration and awe, the more often and steadily we reflect upon them: the starry heavens above me and the moral law within me.' The knowledge of the universe (the starry heavens) and of Good and Evil (the moral law) places each human being in the presence of something grand, incommensurable, of something sublime at the heart of the awareness of one's own existence. Regardless of what this 'sublime' consists of (which depends, for instance, on one's culture, beliefs, etc), it cannot be neglected. Even more than the values or eprinciples that found the ethical approach, it proves that humanity cannot be reduced solely to his animal condition, nor even to the condition described by the name he gave himself, *Homo sapiens sapiens*, literally the human-who-knows-he/she-knows. Kant insisted that this sublime is primordial.

The reference to aesthetics and the call of the sublime must not let ethics give way to the failings of cosmetics. What I mean is that ethics, whether in the form of a committee, an opinion, or an expertise, must not serve to cover up the absence of purpose or the negative consequences of an activity or a program. They may highlight its cultural roots, legitimacy, and advantages, but they must not conceal its limits and defects. Ethics are, or should be, an expression of natural elegance.

# Chapter 3
# A Brief History of Space Ethics

---

'The consequences of a technological conquest can never be predicted' were the words of a French thinker, at the end of the 1950s, writing about innovation. The history of science and technology doesn't offer us any examples to counter this observation. Should all ethical questioning consequently be ruled out of engineering sciences, for fear that it might dampen scientists' and engineers' enthusiasm and prevent further innovations? Does it suffice to hide behind the screen of so-called neutrality and leave it to experts in courtrooms and wise men in ethical committees to define the terms of use of all discoveries, inventions, and innovations? Not so long ago, the answer to those two questions would have been 'yes' (I have previously given examples of this). At the time of their beginnings, space exploration and conquest did not escape this type of questioning.

## The emperor and the philosopher

If I had to name a pioneer in the field of space ethics, I would gladly choose the Chinese emperor Ming Xuanzong, also known by the name of Xuandei. The seventh and last of China's great maritime expeditions heading West took place under his reign. Between 1405 and 1433, the famous admiral Zheng He was in command of a huge armada of seventy vessels and 30,000 men. He explored Java and Malacca, Ceylon and the Maldives, Ormuz and Mecca, as well as Aden and Mogadishu. An extra expedition might have given China the Cape of Good Hope, but the emperor decided otherwise. In 1433, one of his edicts forbade Chinese people to leave their country and ordered the dismantlement of Zheng He's fleet and all the country's shipyards. Building a junk with more than two masts became a crime punishable by death. The Middle Kingdom locked itself up between its walls just like its sovereign confined himself in the Forbidden City, painted with the colors of the sky.

How can we explain Xuandei's decision? It would be easy, and perhaps reasonable, to stamp the 1433 imperial edict with a sense of pride. Why should a Son of the Sky be concerned about territories so distant from the centre of the earthly world, where Beijing lies with its City and its Emperor? Why should his subjects have to leave? Are supreme wisdom and harmony not to be found by his side? Is the Heavenly Empire not brimming with all necessary riches? Unless maybe Xuandei realised that, in 1433, China had started to become too big for him to administer and fragmentation had to be prevented at all costs (building the Chinese armada had required timber from half of Southern China's forests). The emperor may have feared that he would cease to be at the center of his territory if the borders were no longer walls but capes. Maybe he feared he would discover another center of the world and another Son of the Sky . . .

Since then, five centuries have passed. In 1960, three years after the first *Sputnik* (the first artificial satellite) was launched by the Soviets and one year before Yuri Gagarine's flight, Walter Pons published a book called *Steht uns der Himmel offen?* (*Is the sky open to us?*[1]). The German philosopher did not focus on astronautic techniques, on how they should be practiced and what consequences they may have, but rather on the very meaning of space exploration for intelligent and conscious beings like us. To accompany the nascent spatial era, he highlighted a piece of wisdom inspired by the ancient thinkers: 'We will never really get to know the world unless we learn to know ourselves first.' In the wake of Socrates (the Greek philosopher) and Georg Wilhelm Friedrich Hegel (the German philosopher) Pons suggested to make our individual inner conscience the authority of truth and decision-making. There was no longer any question of our being guided by some divine order written in the stars that our observation of the sky would enable us to perceive. Henceforth, it would be a matter of writing our own fate in the stars.

---

1.  Walter Pons, *Steht uns der Himmel offen? Entropie-Ektropie-Ethik. Ein Beitrag zur Philosophie des Weltraumzeitalters* (Wiesbaden: Krausskopf Verlag, 1960).

Neither Xuandei nor Pons ever had the slightest responsibility or even the slightest influence in space exploration. However, each of them in his own way, though for different political or philosophical reasons, was concerned about the soundness and purpose of going beyond the borders of the known world and confronting the unknown. They also tried to foresee the possible consequences of such ventures. This is why I consider these two men pioneers in space ethics.

## The president and the astronomer

John F Kennedy's mindset was not the same as his distant Chinese predecessor's. He had been at the White House for barely five months when he decided to make his country join the race to the Moon. On May 25 1961, he proposed to Congress to 'achieve the goal, before this decade is out, of landing a human on the Moon and returning him safely to the Earth'. The following year, on September 12 1962, this time at the Rice University stadium in Houston, he said 'We choose to go to the Moon. We choose to go to the Moon in this decade and do the other things, not because they are easy, but because they are hard'. He also told his fellow-citizens

> We set sail on this new sea because there is new knowledge to be gained, and new rights to be won, and they must be won and used for the progress of all people. For space science, like nuclear science and all technology, has no conscience of its own. Whether it will become a force for good or ill depends on man, and only if the United States occupies a position of pre-eminence can we help decide whether this new ocean will be a sea of peace or a new terrifying theater of war.[2]

Using such terms to describe the perspectives and challenges associated with the emergence and development of space programs was really a way of elaborating and encouraging an ethical reflection.

The ethical considerations raised by Bernard Lowell in his book called *The Exploration of Outer Space*,[3] published in 1962, broach matters of another sort. After summarizing the available knowledge

---

2. *Public Papers of the Presidents of the United States. 1962*, volume 1, 669–670.
3. Bernard Lovell, *The Exploration of Outer Space* (New York: Harper & Row, 1962).

regarding the conditions necessary for life to occur and develop, he endorsed the old philosophical question about the plurality of worlds—in other words, the possibility of extraterrestrial life. He then mentioned operation West Ford Needles. In 1961 (and again in 1963, when the operation was repeated), the US Air Force released several million copper needles in orbit around the Earth, at an altitude of approximately 2,300 miles. The point was to create a belt of spatial dipoles that could serve as passive reflectors for military communications. Lowell wondered whether it was reasonable to pollute outer space in such a way and for such a reason. He ended his book on a rather optimistic note: the race to the Moon, and more broadly the spatial competition which Americans and Soviets were engaging in, provided excellent opportunities for these two superpowers to devote budgets to activities far less belligerent than the proliferation of weapons of mass destruction.

From the political perspectives opened up by space during the Cold War to the quasi-metaphysical question of extraterrestrial life or the criticism of nascent outer space practices, Kennedy's and Lowell's words enable us to measure the extent of the diversity of ethical questions that can be associated with the development of space activities.

## Fruitless institutional initiatives

Twenty years later, in 1982, the second UNISPACE conference took place in Vienna. Its mission was to pursue and enrich the ongoing international reflections about space-related politics and laws. Fifteen years after the space treaty, several delegates were worried that the militarization process was still threatening space and that developing countries were not benefitting enough from space technologies. They called for more cooperation and solidarity in space as well as on Earth. The Apollo missions and the solar system exploration missions had shown the singular position occupied by our planet as well as its limits. Could space not offer the opportunity of conquering and colonizing new territories to the benefit of all humankind?

The following year, a similar spirit inspired an initiative on behalf of the United Nations' Educational, Scientific, and Cultural Organization (UNESCO). The Paris-based international organization put VS Vereschtin (vice-president of the Interkosmos Council under the USSR Academy of Sciences) in charge of preparing a meeting about international cooperation in space. Vereschtin was convinced that 'Preserving space as a haven of peace and cooperation between the world's nations, and not allowing humanity to get accustomed to the idea that militarization of space is supposedly inevitable, is one of the chief objectives of law and ethics at the present time'.

On December 16, 1983, a teleconference brought together six participants from four different continents. The Austrian Peter Jankowitsch, who was then the president of the Committee on the Peaceful Uses of Outer Space (COPUOS), remarked with satisfaction that space was giving humankind a new vision of itself and was thereby leading to the elaboration of a new set of ethics. That being said, he lamented that

> Whilst the first decades of the exploitation and use of space were characterised by a fortunate evolution of principles [of non-acquisition of rights of sovereignty in space and of celestial bodies, exclusively peaceful use in the interests of mankind] and whilst in particular the cooperation between the main space powers, the Soviet Union and the United States, have made great headway—the joint flight baptised *Apollo-Soyuz* constituting a spectacular step in this process— the most recent years in space history are somewhat lacking in examples of international cooperation. On the contrary, several of the main principles pertaining to the use of space— and in particular the principle of an exclusively peaceful use of space activities—appear increasingly threatened.

Throughout this exchange, the American Isaac Rasool showed more interest in the ethical aspect. He evoked the various kinds of pollution in space, the status of the data and information gathered, etc. Establishing a connection between ethics and international law, the Czech Vladimir Kopal underlined the need to define a general principle of governance alongside the declaration of space as common heritage of humanity.

This UNESCO initiative was remarkable in more than one way. It gauged space activities in the light of the space treaty, proposed fifteen years earlier. It raised the main ethical questions and introduced the issue of governance, inherent to the respect of the principles of space law. It also helped to define a possible perspective for space ethics in close connection with the development of laws. Unfortunately, this initiative had no direct result. It took the UN organisation another twenty years to show interest in space again.

A third institutional initiative took place in Casablanca in March 1984. The Academy of the Kingdom of Morocco organised a conference under the title 'The deontology of the conquest of space'. It came within the scope of the spirit of the UN initiative. If most speakers showed concern for the threat of increased militarization of space (President Reagan had launched his Strategic Defense Initiative, often tagged as 'Star Wars', one year earlier, in March 1983), they were also aware of the other challenges raised by space activities; namely those regarding technical expertise (cluttering of orbits, proliferation of debris, pollution caused by launchers, etc.) and those pertaining to judicial and diplomatic matters (for example, whose sovereignty? what data sharing?).

The Casablanca participants highlighted the fact that one of the main challenges that the international and space communities had to address was not so much to revive and refine the spirit of Vienna, the Space Treaty, and the international agreements previously written up and signed, but to effectively put them into practice. In other words, to elaborate an explicit deontology and apply it. How could this tour de force be achieved without coercion, resorting to nothing but good will? One of the speakers at this meeting wondered whether the 'damned of the earth' wouldn't soon be joined by the 'damned of space', due to the difficulties or even the impossibility faced by Third-world countries to access data and, even more so, space technologies. How could they not feel excluded and dispossessed when confronted with the arrogance of the great space-faring nations and the prevailing disparity between the different players on the world scene? Lastly, the Casablanca conference evoked the consequences of the use of space assets on cultures and on their dissemination or, on the contrary, their isolation and eventual disappearance. Is it possible to think or to claim that space has already enabled—and might continue to enable—true innovations in sociocultural matters? Thirty-five years later, this question remains relevant and does not yet seem to have received a satisfactory answer.

Although they remained fruitless, these three initiatives show that, at the beginning of the 1980s, the international space community was driven by a concern for ethical and deontological questions. The persistent threat of militarization of space and the need to see space laws evolve were certainly causes of this concern. Unfortunately, notwith-standing some speeches and acts published by the Moroccan institu-tion, nothing remains of these three initiatives. In the second half of the 1990s, the Japanese space agency NASDA initiated a study on the cultural consequences of space activities on the Japanese society, but in spite of this, fifteen more years had to go by before ethics returned to a more prominent and solid position at the heart of the space community.

## Two space agencies commit

One has to admit that, compared with genetic engineering or nuclear technology, space is not scary enough to awaken any particular ethical upheaval on behalf of political leaders or civil societies. This remained true when, in February 1986, the Challenger space shuttle crashed off the North American coast with its seven crew members onboard and when, seven years later, the Columbia space shuttle burst into flames in the Earth's atmosphere. Following each of these two catastrophes, the American authorities named a commission of inquiry but never an ethics committee. Ultimately, it was in Europe that ethics came to the front of the scene in the space sector.

'At the initiative of the Director-General of UNESCO, Mr. Fed-erico Mayor, and acting on a proposal by the Director-General of the European Space Agency (ESA), Mr. Antonio Rodotá, a new working group was set up to consider the ethics of outer space in December 1998 on the basis of a partnership between UNESCO and ESA.' These were the terms used in the report published by UNESCO in July 2000, under the title *The Ethics of Space Policy,* to describe the roots of the space community's renewed interest in ethical matters: an ESA proposal and a UNESCO initiative. Coordinated by Professor Alain Pompidou, the working group sought 'to identify the difficulties and fears, opportunities and promises associated with the conquest of space, while providing the necessary explanations in the clearest and most comprehensive manner possible, taking into account the needs of the populations in their specific sociocultural context'. To the two components usually included in the definition of space (1. a dimen-sion, a place, an environment, and 2. a tool, a technical ensemble), the report explicitly added a third; namely, the general public's perception within a specific social and cultural reality. Taking into account the social dimension (of space, in this case) was an essential contribution to the ethical process overall. In contrast, although military activities had been at the center of concern and ethical questioning during the previous decade, they were absent from the report requested by the ESA from UNESCO. This was probably the case because the ESA's projects and programs statutorily belong exclusively to the civilian sector and do not contribute to the weaponisation of space or to space dominance.

In the wake of this report, at the heart of its World Commission on the Ethics of Scientific Knowledge and Technology (COMEST[4]), UNESCO created a subcommission dedicated to outer space. But after having led several actions, most often with the support of the ESA (reports, conferences, etc), this subcommission became dormant from 2005 onwards. One of the reasons for this disinterest in space by UNESCO may have been the result of redundancy or competition with another UN organization called COPUOS. It was probably no coincidence that, in June 2001, after the COMEST published its report on *The Ethics of Space Policy,* COPUOS dedicated one of the sessions at its annual meeting in Vienna to the theme of space ethics: an obvious way to 'mark its territory' and remind everyone that COPUOS was the first to show interest in space ethics and dedicate itself to this matter, due to its judicial competence.

Twenty years after the ESA's initiative and regardless of the competition or complementarity between these two UN organizations, both of them gave up all interest in space ethics. And the ESA did not follow the example of its French counterpart, the *Centre national d'études spatiales* (CNES).

Indeed, in early 1999, Gérard Brachet, the executive director of the CNES, entrusted a group of engineers with the task of laying the foundations of space ethics. Two years later, the board of the CNES created the position of Ethics Adviser, the first of its kind in the world of astronautics (and the only one to this day). Published in October 2001, *La Seconde Chance d'Icare* (*Icarus's second chance*) was the result of these first three years of reflection.[5] CNES' research on matters of space ethics was developed throughout several publications. It

---

4.  Acronym taken from the French name «Commission mondiale d'éthique des connaissances scientifiques et des technologies».

5.  *Cf* Jacques Arnould, *La Seconde Chance d'Icare. Pour une éthique de l'espace.* (Paris: Cerf, 2001), and *Icarus' Second Chance. The Basis and Perspectives of Space Ethics* (Wien/New York: Springer, 2011).

was also presented to various audiences at colloquiums and conferences open to the general public, and it occupies a regular column in the news magazine of the French space agency. Several of the CNES' groups of expertise and foresight have introduced an ethical perspective into their work.

## A relevant distinction

It is fitting to say that, thanks to the ESA, UNESCO and CNES, ethics have entered the spatial world in a durable way. For example, conferences organised by the International Astronautical Federation (IAF), the International Academy of Astronautics (IAA), and the Committee on Space Research (COSPAR) welcome communications on the topic of space and ethics. The International Space University (ISU), established in Strasbourg (France), has honored its intercultural dimension by opening a summer school class and a masters degree program on ethics. The European Science Foundation (ESF), the European Space Policy Institute (ESPI), which study the future of space activities, have also integrated this new field. The new European Astrobiology Institute (EAI) has introduced it into its research. No space agency other than the CNES, however, has designated a person or a team in charge of surveilling the ethical dimension of its activities. Until now, no ethics committee in charge of space-related questions has been created anywhere, at any level. This is not surprising. On one hand, the pressure exerted by public opinion, current events, and the near future is weak or even nonexistent. On the other hand, the multiplicity of fields involved would probably complicate its constitution and management. In short, for most space-related organizations (and even more so for the academic world[6]), ethics remain a frontier that has yet to be crossed.

Nonetheless, we mustn't lose faith. As John F Kennedy had done in the early 1960s, the political world can still decide to support an ethical approach in the space sector. An American publication confirmed this in 2009.

---

6. There are, however, some noteworthy exceptions: Eugene C Hargrove, *Beyond Spaceship Earth. Environmental Ethics and the Solar System* (San Francisco, Sierra Club Books, 1986); Tony Milligan, *Nobody Owns the Moon. The Ethics of Space Exploration* (Jefferson: McFarland, 2014); James Schwartz and Tony Milligan (editors), *The Ethics of Space Exploration* (New York: Springer, 2016).

On February 1 2010, Barack Obama announced the cancellation of the Constellation program, which had been created by his predecessor, George W Bush, in 2004. The goal of this program was to send astronauts to the Moon around 2020 for long-term missions. Regardless of how this presidential decision was eventually achieved (in fact, Constellation was never completely cancelled . . .), three reasons were given to justify it: an excessive overrun of the budget allocated to the program, the failure to comply with deadlines, and the absence of true innovations associated with the project. It was a hard blow for many players of the American space sector, but not for all. The report, written by a commission directed by Norman Augustine, published in October 2009 under the title *Review of United States Human Space Flight Plans*, deemed that NASA should henceforth rely more on private operators for low-orbit activities. Once implemented, this orientation contributed if not to the emergence, at least to the acceleration of the development of NewSpace, in other words of the new spatial entrepreneurs, thanks to the reinforcement of their partnership with NASA. This report contains another noteworthy peculiarity: it is perhaps the first official document (in the common sense of the term) that, in its own way and fashion, mentions and even promotes an ethical approach applicable to space. Let me explain.

At least on two occasions, this text proposes simple and efficient criteria to evaluate space programs and even to make decisions.

> We explore to reach goals, not destinations. It is in the definition of our goals that decision-making for human spaceflight should begin. With goals established, questions about destinations, exploration strategies, and transportation architectures can follow in a logical order. While there are certainly some aspects of the transportation system that are common to all exploration missions (for example, crew access and heavy lift to low-Earth orbit), there is a danger of choosing destinations and architectures first. This runs the risk of getting stuck at a destination without a clear understanding of why it was chosen, which in turn can lead to uncertainty about when it is time to move on.[7]

---

7. *Review of US Human Space Flight Plans Committee;* Austin Augustine; Kennel Chyba; Crawley Bejmuk; Chiao Lyles; Greason Ride, http://www.nasa.gov/pdf/396093main_HSF_Cmte_FinalReport.pdf, 33.

It is not difficult to apply this method to the Constellation program and, therefore, to understand the American government's decision. The destination of this program was obvious: the Moon. Its goals, however, were not defined well enough. Therefore, Augustine and the members of his committee concluded that it would be best to suspend, cancel, or at least seriously reconsider Constellation. Which is what President Obama decided to do in early 2010, no matter what the final effects of his decision would be.

Without glossing over it, the frame of mind promoted by Augustine's report is interesting. Rather than distrust or criticise *a priori*, rather than come up with sterile 'What's the point?' reactions, it is better to try to discern the deep motives behind an enterprise, its finalities (may they be goals and/or destinations), and to assess its consequences, whether expected or feared. This frame of mind is all the more relevant since, nowadays, and in particular within the space community, the engineers' conviction 'We can do it!' prevails and underlines a performance-driven culture. This culture requires enthusiasm, tenacity, courage, and an ability to accept risks, but it can easily confuse destination and goal. In fact, this leads us to a sensitive issue that must be raised in regard to space activities and technological development in general, a tricky question which seems ever harder to answer; namely, does the function create the system, or does the system create the function? At this stage, it becomes necessary to step back, put things into perspective, and look at them under a different a light. Then, to debate and disagree, and, last but not least, be convinced and make the right decisions. This process has nothing to do with a hobby for bored citizens or an alibi for careful leaders. It involves nothing but the indispensable awareness of our actions, at both the individual and collective levels. Nothing other than what we call 'ethical questioning'.

# Chapter 4
# Law is a Mirror of Space Ethics

---

At a time when NewSpace is starting to look like a new American frontier or like a new gold rush, one question has become particularly relevant: is space for sale? Or, more generally speaking, is space devoid of any judicial framework, of any rules or code of good conduct? Those who are selling plots of land on the Moon or on Mars are casting doubt in the minds of our contemporaries who, curious or concerned about the future of space, usually don't know that there is an entire legal corpus aimed at defining, orienting, or even inspiring space-related activities. In a nutshell, space is not above the law.

I have no intention here to present the contents of space law, its applications and limits, nor to describe the international structures responsible for its elaboration, implementation, and evolution. In this chapter, I merely wish to clarify what inspired its genesis and what lies at its foundations. For if there has never been any shortage of debates about the place a legal framework should be given in ethical matters, I believe it is possible, and even necessary, to associate law and ethics in the astronautic field. In other words, it is fitting to identify space law as the oldest ethical expression with regard to modern astronautic activities, as well as one of the key sources providing our contemporaries with food for thought.

## The sword and the bench

Between 1882 and 1883, the International Polar Year took place. Alongside expeditions led under national flags (first and foremost, the United Kingdom and Norway), this initiative brought together researchers from various countries in order to carry out joint scientific missions destined to explore the Earth's polar regions.

Established from July 1 1957, to December 31 1958, the International Geophysical Year (IGY) was organised along the same lines. After World War II and within the context of the Cold War, the scientific community wanted to promote cooperation between countries, especially in terms of sharing knowledge about our planet. Over 60,000 scientists and engineers from sixty-six countries took part in this initiative, the main goal of which was to study global phenomena occurring on Earth or in its direct spatial environment: gravity, geomagnetism, physics of the ionosphere, meteorology, oceanography, seismology, Northern lights, solar activity, cosmic rays, etc. The astronautic dimension of this initiative became clear in summer 1955, when the United States and the Soviet Union successively announced their intentions to launch satellites on the occasion of the IGY. Sputnik 1 was launched on October 4 1957, and Explorer 1 on January 31 1958 (leading to the discovery of the Van Allen radiation belts). The IGY also gave countries such as France, the United Kingdom, Japan, Canada, and Australia the opportunity to develop sounding rocket programs to explore the upper atmosphere. To use an expression suggested by Roger-Marie Bonnet, ex-director of scientific programs at the ESA, this was the role played by the 'bench' (that is, the scientific community) in the modern history of space. This role was characterised by a willingness to engage in research, to explore and . . . to cooperate. Conversely, the role of the 'sword' (that is, the military) seems just as important.

Beyond the field of ballistic missiles (inaugurated by the development of the German V-2 rocket), and following the example of aerial history, the military quickly grasped the benefits it could derive from space technology. Six weeks after the success of the first Soviet Sputnik, an American general warned his fellow-citizens in very clear terms:

> Whoever is capable of controlling the airspace can also control the lands and seas below it. I believe that, in the future, those who will have control over space will also be able to take control on the surface of the Earth. Regarding the sky and space, I wish to point out that there is no actual separation between the two. They form an indivisible field of operations.[1]

---

1.  *Cf* Paul B Stares, *Space Weapons and US Strategy. Origins and Development* (London/Sydney: Croom Helm, 1985).

And four days after the launch of the first American satellite, American senator Hubert H Humphrey was pleased to point out the recovered balance between the two superpowers. He announced a new era, opened by the upcoming, predictable arrival of reconnaissance satellites: 'A satellite of this type would impress all nations of the world by proving that countries and national borders are no longer sacrosanct. Due to its very essence, it would provide a striking example of internationalism, implying the creation of new concepts of international law and a new international order'.[2]

Thus, space law developed in a context of international scientific cooperation as well as strategic competition between the United States and the Soviet Union.

## Extraterrestrial *versus* terrestrial

'*Cuius est solum, eius est usque ad coelum et ad inferos*—Whoever's is the soil, it is theirs all the way to Heaven and all the way to Hell.' This is a Roman principle, consolidated and commented on in the Middle Ages by the jurist Accursius. If it were applied to outer space, it would entail that a state's property, and consequently its sovereignty, is maintained without limits in space (let's leave aside the matter of 'Hell'). Yet we know that, among the principles of space law, those of freedom and non-appropriation are at the top of the list. So, how are we to deal with this paradox?

The precursors and pioneers of space law raised this question from the outset: should this field be considered as an extension of air law? The Belgian Emile Lande, as early as 1910, and the Czech Vladimir Handl, in 1932, argued that space flight was something specific and, therefore, required a new and independent set of laws. In contrast, in 1934, the Soviet Evgeny Korovin defended the idea of merely extending air law, maintaining states' rights to sovereignty and self-defense in outer space. This debate between jurists continued and evolved until the 1950s, when it focussed on the following question: should legal regulation depend on the location and nature of the activity at issue? In other words, when applied to space, the question became: is the sovereignty of a state delimited by a topographic definition or a functional definition? Here again, the options were to define space

---

2.  *Cf* Philip J Klass, *Secret Sentries in Space* (New York: Random House, 1971), 220.

either as a place or as a set of tools and activities. In the end, space law developed like a specific law, which rested on a definition and delimitation of space, precisely described as *outer* space, and which remained connected with maritime law and air law.

Far from being anecdotal, the debates raging at the time of space law's emergence illustrate the essential bond that can and must connect judicial matters and ethics. Contrary to common beliefs, the main purpose of each of these is not to protect and maintain the status quo, but rather to be applied whenever this status quo is questioned, regardless of why or how. It is when it comes to crossing a border, or triggering or facing a revolution, that rules and laws are really necessary, just like ethical practices or approaches. As the pioneers of space law (Lande, Mandl and Korovin) had understood, it isn't easy to measure to what extent a science or technology is novel, or to determine the consequences of its applications. It is sometimes necessary to invent new definitions and impose new conditions and new rules, without omitting to reiterate what the known goals are, to define them more precisely, and to set new ones.

In addition, if the emergence of a new technological field and the concomitant exploration of a new geographic area can lead to a new legal corpus, the consequences of such changes on the 'old' fields may not be neglected: extraterrestrial matters must not overshadow terrestrial concerns. Although one of the first questions raised by jurists relates to the relevance of maintaining the states' sovereignty in space, one must also keep in mind that space activities influence the notion of sovereignty on Earth, its respect, and its management. Scientists' enthusiasm to collaborate during the IGY cannot and must not eclipse the spirit of competition that used to drive the relationships between the United States and the Soviet Union, the spirit of the race to the Moon.

### *In principio*

The United Nations General Assembly didn't wait long before it broached the subject of outer space. On November 14 1957, barely six weeks after the launch of Sputnik 1, the UN issued a resolution recommending a peaceful use of space. Two more resolutions followed, one in December 1958 and another in December 1960. These led, in December 1963, to the signing of the 'Declaration of legal principles governing the activities of states in the exploration and use of outer space'. This paved the way for negotiation and resulted in the writing of a general treaty and supplementary agreements. Several more resolutions were made between 1982 and 1996.[3]

---

### Inter-state treaties and agreements

* 'Outer Space Treaty'—*Treaty on Principles Governing the Activities of States in the Exploration and Use of Outer Space, including the Moon and Other Celestial Bodies* of December 19, 1966, opened for signing on January 27 1967, entered into force on October 10, 1967. It was ratified by 98 states and signed by twenty-seven more. Its principles were then completed and developed by other international texts:

* 'Rescue Agreement'—*The Agreement on the Rescue of Astronauts, the Return of Astronauts and the Return of Objects Launched into Outer Space* of April 22 1968, entered into force on December 3 1968, ratified by 90 states;

* 'Liability Convention'—*The Convention on International Liability for Damage Caused by Space Objects* of March 29 1972, entered into force on September 1 1972, ratified by eighty-six states;

* 'Registration Convention'—*The Convention on Registration of Objects Launched into Outer Space* of January 14 1975, entered into force on September 15, 1976, ratified by fifty-one states;

* 'Moon Treaty'—*The Agreement Governing the Activities of States on the Moon and Other Celestial Bodies* of December 18 1979, entered into force on December 18 1984, ratified by thirteen states.

---

3.   Resolution of December 10 1992, on direct television broadcasting; resolution of December 4 1986, on remote sensing; resolution of December 14 1992, regarding nuclear sources in outer space; resolution of October 28 1992, on cooperation in favor of developing countries. *Cf Text and status of treaties and principles governing the activities of States in the exploration and use of outer space, adopted by the United Nations General Assembly A commemorative edition, published on the occasion of the Third United Nations Conference on the Exploration and Peaceful Uses of Outer Space (UNISPACE III)* (Vienna: UN, 1999).

As we know, the perspectives currently being opened in terms of space use and exploration, especially those advocated by NewSpace stakeholders, call into question and maybe even endanger this legal corpus. To deal with this challenge as well as possible, it is indispensable to repeat and clarify the principles that inspired these fundamental texts. Any possible evolution or revolution will necessarily take place according to these rules, or somehow in reference to them. Either they will be taken into account, or they will be forgotten or ignored. Principles should be considered as vehicles carrying practical solutions, adapted to each situation and each specific case. Full of positive values, they introduce a moral dimension into the realm of law. In short, principles are the bonds between ethics and law.

The history of law shows that treaties concerning international spaces established particularly structuring principles. Because the sea, the sky, and outer space are interstate binding factors or, on the contrary, isolating factors, the principles on which such treaties are based contribute to a sort of universal order. The main idea is to oppose the constitution of exclusive sovereign rights and to establish competing authorities between states. In other words, competition is preferred to exclusivism, that is *imperium* (political power) should win over *dominium* (property). Freedom of access and non-appropriation are the two key principles here. They inspired all the other norms of the legal corpus, which was founded on six basic principles in total: freedom, non-appropriation, states' international responsibility, peaceful use, cooperation, and respect for the common interest.

Some authors consider these principles as the 'mantle of respectability of space law'. This critique must be taken seriously. Indeed, people sometimes apply the law or argue in favor of ethics in order to wipe their consciences clean, or even to divert attention towards other fields. Though this possible use of law and ethics cannot be denied, it shouldn't serve as a pretext to ignore or set aside the 'spirit of laws', in the original sense of 'principles'—quite on the contrary. For example, the notion of freedom of access to space is not the sum of specific liberties that could potentially be claimed, but a founding norm (when the treaty was elaborated in 1967, space technologies were still in their infancy). Such freedom calls for responsibility and involves cooperation, solidarity, and equity. Juridical standards thus represent an ethical gauge to evaluate space-based activities. Similarly, the principle of non-appropriation cannot be reduced to merely prohibiting acquisition. Rather, it defends the simultaneous coexistence of several authorities, political lines, and ranges of actions. Whether we are dealing with circumterrestrial orbits and frequencies, which are both limited resources, or with celestial bodies and the exploitation of their resources, the principle of non-appropriation is not easy to apply, because dynamics of appropriation are being established in space. Consequently, it is indispensable to set up a proactive governance in regards to this principle.

## New limits

During the international astronautics congress that took place in Jerusalem, in October 2015, a session dedicated to space law experienced a moment of liveliness, or even of unrest. This occurred after Professor Henry Hetzfeld's presentation called 'How simple terms mislead us: the pitfalls of thinking about outer space as a commons'. The American jurist at George Washington University openly questioned the meaning, and especially the application to space, of the concepts of common interest, common heritage, and province of all humankind, even though these are the very foundations of international space law. A few weeks later, on November 25 2015, the White House announced that Barack Obama had signed the Space Resource Exploration and Utilization Act, a text proposed by the House of Representatives in May of that year. The aim of this act was to clarify the legal framework around the property rights to resources obtained by exploiting asteroids and, eventually, to facilitate and encourage the commercial exploitation of these resources by private companies based in the United States. After showing interest in the fields of telecommunications, Earth observation, and, more recently, spatial transportation and suborbital flights, the American administration came up with this new initiative in favor of private players' commitment to the commercial use of space, thereby providing an efficient way to support and preserve American leadership and supremacy in the spatial realm. On February 3 2016, Etienne Schneider, vice-prime minister and minister of the economy of the Grand Duchy of Luxembourg, revealed a series of measures destined to attract to his country companies specialised in exploiting mineral wealth from space.[4]

---

4.  *Cf* Jacques Arnould, *Oublier la Terre? La conquête spatiale 2.0* (Paris: Éditions Le Pommier, 2018).

With these US and Luxembourg initiatives, NewSpace truly entered the political and judicial spheres. Henceforth, jurists were forced to measure the solidity and relevance of the foundations of space law, its principles, and its capacity to take into account the emergence of new players, new tools, new resources, and new markets. Indeed, if we focus on mining operations in space, it is not easy to come up with an answer, let alone a legal corpus, that simultaneously takes into consideration the current space legislation, its possible evolution and transformation, the diversity of mining laws and practices, and their application to the particularities of space, as well as all the existing sociopolitical constraints and perspectives.

NewSpace, however, is still only in its infancy. Therefore, we needn't be concerned about the legal possibilities and limits that should accompany its future development before we seriously address the ethical questions and political aims associated with it. I do not intend here to contest or criticise the efforts of those who are giving thought to the conditions in which NewSpace will develop, particularly in the field of law. Indeed, it is rare to see legal consideration develop alongside a technological, socioeconomic evolution, or even precede it, rather than try to 'catch up' with it. Nonetheless, this promising situation may not lead us to ignore the flaws and limits of current work on the subject.

I recognise how competent the jurists are who get together to form the working group called *The Hague Space Resources*, and I welcome their commitment to this joint initiative of the Dutch Ministry of Foreign Affairs and the University of Leiden[5]. Originally coming from academic environments or NewSpace companies, and representing various space-faring countries, they work at elaborating building blocks, foundations meant to serve NewSpace's organisation and legal management in a near future. They clarify what terms such as 'space resource' or 'operator' actually refer to; they update the principles of space law and their current interpretation; they broach questions of security, of responsibility in case of damage, and the tricky subject of sharing benefits; and they imagine how an adaptive governance could emerge. Yet, in addition to pondering how international institutions such as the COPUOS might react to their work (a question which

---

5.  *Cf*    https://www.universiteitleiden.nl/en/law/institute-of-public-law/institute-for-air-space-law/the-hague-space-resources-governance-working-group.

they certainly ask themselves, too), I wonder whether their reflexions aren't limited and crippled by the fact that they stick to a representation of the world that is now outdated. To be more precise, I wonder whether these questions aren't indicators of how the world's representation underlying space law has failed to adapt to present realities. Let me explain.

The days are over when the leaders of Spain and Portugal could split the cake of the New World and of the entirety of the known world by tracing meridians on maps. We no longer live at a time when companies and trading posts could settle on distant coasts and lands, with the self-serving authorization of their sovereigns. As vast as it may have seemed in those days, the world remained limited, as our predecessors wound up noticing. Nowadays, our world, that is the world of the space era, no longer has such boundaries. Of course, the limits immediately associated with our planet haven't changed. In fact, they are even felt more and more tragically as we increasingly fear the permanent depletion of our terrestrial resources. That being said, we can still dream of colonizing other planets, of finding and exploiting elsewhere what will be necessary for our survival and even for our cosmic expansion. Henceforth, why should we not continue using yesterday's dreams, for example those of science-fiction authors, to construct tomorrow's reality? In that case, there is no reason to discern any true limit to our enterprises. Even possible others, 'aliens', 'alter egos', remain at distances so great that we struggle to imagine how they could possibly impose their limits on us. In other words, we remain among humans (or among inhabitants of the Earth, if we wish to take other living beings into account), forced to elaborate our own rights and duties towards each other, but without any external limits to rely on. Like for space miners, the sky no longer seems to be a limit. And law, in its current state, doesn't offer sufficient protection to fully escape a certain sense of vertigo.

Yet, vertigo can be beneficial and invite us to raise truly ethical questions. Why not find inspiration in this dizzying experience, this 'overview effect' as described by astronauts in orbit around our planet?

# Chapter 5
# Piecing Together the Earth

Until the end of the eighteenth century, only the more spiritual humans, the scholars and poets, had dared climb the steps of the mystical experience, patiently scrutinizing the stars and elaborating the boldest theories, giving free rein to their creative imagination in order to escape their earthly prison and manage to look down upon it. People had to wait for the invention of lighter-than-air and heavier-than-air vehicles to gain bodily access to the third dimension, thus seeing the Earth from above for the first time, through instruments or with the naked eye. When reality finally met imagination and fiction, it revolutionised humanity's bond to its planet and to its environment.

The history of the observation of the Earth from above was marked by the flight of Explorer 2, a manned high-altitude balloon which, in November 1935, reached a record altitude of 72,395 feet (22,066 meters) and took the first oblique photograph of the layers of the atmosphere. The Earth's curvature was visually confirmed, as well as the thinness of its atmosphere. At the same time, the immense and empty dark space surrounding our planet became visible. After World War II and in the context of the Cold War, thanks to instruments embarked onboard the German V-2 rockets that returned to the United States, it was once more the study of the atmosphere and of its penetration by guided missiles that resulted in the 'by-product' of space photographs of portions of the Earth. Henceforth, they could be put together like a mosaic to form a global view. That was when the idea emerged to use meteorological satellites to acquire a synoptic view of our planet. The first meteorological satellite, Tiros-1, was launched on April 1, 1960. This 'patrol vehicle of hurricanes and storms' not only transmitted images of clouds, obtained through cathode ray tubes, but also images of hydrologic, oceanographic, and terrestrial phenomena. At the beginning of the 1960s, the observation of Earth from space started to interest scientists, politicians, economists, and military experts. Besides, as if to convince public opinion of the utility of spatial technologies, the image of the Earth was shown more and more frequently on the first television screens.

## At last, the Earth as a whole!

For Alfred Sauvy, a renowned French economist and sociologist, there was no doubt that 'Man's walking on the Moon was at the origin of the contemporary ecological movement'.[1] This idea may be flattering for the Apollo program and for the space enterprise in general, but it is incorrect. Even if it is tempting to try to establish a causal link between space and the concern for the state of our planet, thus finding an additional argument in favor of the development of space technologies, such a correlation does not exist. It would be more appropriate to speak of a coincidence or concomitance. Indeed, in 1962, American scientist Rachel Carson published *Silent Spring*, a book in which she denounced the harmful effects of DDT and other pesticides on human and animal health, as well as on the environment. Her best-seller resulted in the creation by the United States of the Environmental Protection Agency. This symbolically marked the beginning of the ecologist movement. The first World Conference on the Environment took place ten years later, in 1972, in Stockholm, six months before the Apollo 17 mission, which was the last of the American lunar program. So, even though it is incorrect to defend the idea of causality rather than concomitance, there is no doubt that the photographs of the Earth taken during the lunar missions were quickly taken up by ecological movements and used as icons. Interestingly, they owe a major part of their success not to a scientist or ecological activist, but to an artist of the counter-culture.

In the mid-1960s, 'The Company of Us' (or USCO) brought together American artists who created immersive performances and psychedelic experiences, thanks to the most modern technologies or to the use of LSD. With their motto 'We Are All One', they made it to the front page of *Life* magazine in September 1966. Steward Brand

---

1.  *Cf* Alfred Sauvy, Préface à *L'État de la planète* (Paris: Éditions Economica, 1990).

used to frequent this counter-culture group. One day in February 1966, on the roof of a house at North Beach, San Francisco, he had somewhat of a revelation after taking LSD. He later explained having wondered 'How could I incite NASA or the Russians to turn their cameras around? We should make a button! A badge with a request: "Take a photograph of the entire Earth". No, we have to come up with a question that will speak to America's tremendous paranoia: "Why haven't we seen a photograph of the whole Earth yet?" And that was it. The next morning, I was busy printing buttons and posters bearing this question.' Brand sent his production to NASA, to American Congressmen, United Nations diplomats, and Soviet scientists. On the campus of the University of California, in Berkeley, he sold his buttons for twenty-five cents. Then he offered them to students of Stanford, Columbia, Harvard, MIT, etc.

A few months later, on August 23, 1966, the first 'earthrise' reached us from the Moon's orbit. It was an image taken by the space probe Lunar Orbiter 1, which NASA and Boeing engineers ordered to turn around towards the Earth just before it disappeared behind the Moon. In truth, this operation had not been planned in the mission's official protocol. As soon as it was broadcast, this photograph quickly became known as the 'Picture of the Century'. Was Brand's petition responsible for this turning around? Actually, the picture provided by Lunar Orbiter did not offer a view of the Earth in its entirety. One would have to wait for the Apollo missions, and even for the last one of those, to see Brand's wish fully come true. On December 7, 1972, the astronauts of the Apollo 17 mission took various photographs of the Earth on their way back to it. The most famous one of them was called 'The Blue Marble'. It soon stole the thunder of the 1966 photo, only sharing its success with 'Earthrise', a photo taken by the astronauts of Apollo 8, on December 24, 1968. Thirty years earlier, Martin Heidegger had prophesied that it wouldn't be long before humankind entered 'The Age of the World Picture' (*Die Zeit des Weltbildes*).

Steward Brand's success story doesn't stop here. To USCO's 'We Are All One' motto, Brand added another catchphrase, a new idea: 'Do It Yourself'. He explained 'We are like gods and we must become good at doing things'. So, he undertook a new action; namely the publication of a magazine that would bear an image of the whole Earth and the title 'Whole Earth Catalog'. It was to be a catalog of 'tools' (books, maps, newspapers; tools for gardening, carpentry, and masonry; tents, hiking shoes, kayaks, dinghies, etc) aimed at developing 'people's power to take care of their own education, find their own inspiration, shape their own environment, and share their adventures with anyone who might be interested'. In 2005, ten years after the last copy of Brand's catalog was published, Steve Jobs recalled 'When I was young, there was this amazing publication called 'Whole Earth Catalog', which was like a bible for my generation. It was a bit like Google on paper, thirty-five years before Google's existence. It was an idealist review, full of marvelous tools and awesome ideas'. In 1985, Brand came up with an electronic version of his magazine called the WELL, The Whole Earth Lectronic Link. It was a system of teleconference based on a central computer from which people could communicate in real time. Thus, the American artist embodied the link between the view of the Earth and the emergence of a global human consciousness, a 'noosphere', according to the term tagged by Vladimir Vernadsky. This was no minor effect of the space enterprise.

## Overview effect

Mission Proxima took place from November 17 2016, to June 2 2017, onboard the International Space Station (ISS). Thomas Pesquet, one of the astronauts embarked on this mission, shared with the French public all the enthusiasm, emotion, and concern that filled him upon seeing the Earth from above. 'My perspective changed. What I saw made me want to tell people to do more for the environment. Astronauts have the chance to see how fragile the planet is, to get a global point of view.' He concluded 'If I can play a role as a citizen of the world to encourage people to get involved, I will'.

The French astronaut's experience is not one of a kind: it is shared by all the men and women who travel in the Earth's orbit and by those who made the voyage from the Earth to the Moon. Frank White described it as the 'overview effect'.[2] This experience is probably one of the most unexpected consequences of manned flights. Whoever is freed from terrestrial gravity and undertakes a voyage toward the stars doesn't forget the Earth where he is from, his 'cradle', to use Tsiolkovsky's famous expression. On the contrary, he admires its incred-

---

2.   *Cf* Frank White, *The Overview Effect. Space Exploration and Human Evolution* (Reston, VA: American Institute of Aeronautics and Astronautics, 1998).

ible beauty and discovers its alarming fragility. The overview effect is a singular thing and must lead us to raise a few questions.

First, we should question the words that accompany the pictures taken by astronauts or obtained via observation satellites. Let's simply take the example of the claim, so often repeated, that there are no borders visible from space. This is not true. Without even mentioning the natural borders that separate people and nations (how could one deny the existence of the Himalayas or the Atlantic Ocean?), artificial lines are also perfectly visible from a space shuttle or an orbital station. For example, the separation between North and South Korea can be seen during the day, and even more so at night, as described by the South-Korean astronaut Yi So-yeon. It is undeniable that words can weigh just as much as photos, and be just as dramatic.

In Berlin, in 1933, on the occasion of the opening of *Die Kamera*, an exhibit dedicated to photography, Joseph Goebbels shared his opinion as a war lord: 'We believe in the objectivity of the camera and are skeptical of anything mediated orally or in type'.[3] The Nazi leader used to claim that he preferred objective photos to observations made by humans embarked on aircrafts or to pilots' radio transmissions. Given the specifications of modern sensors, it can be very tempting to agree with Goebbels and to rely on technical, mechanical objectivity, based today on algorithms, pixels, computer instructions, and simulations. But does the Earth thereby not risk losing its soul, and humans a part of theirs as well? Photos of the Earth taken from space are pure products of technology, now readily available on our computer screens for daily, more or less constructive, often commercial use. And in spite of this, people continue to marvel at the photos taken by astronauts through the portholes of the ISS, commented on, and massively displayed on social networks. How can one explain such passion and fascination, even though none of these sights is new or original for anyone who is slightly educated in space matters, or has Internet access or books in his library that were published in the wake of manned missions? How can one justify such great interest by other than the fact that, contrary to the Nazi propaganda minister's opinion, humans need both words and photos, written and oral messages, which can be conveyed by astronauts, poetic geographers, and

---

3. *Cf* Monique Sicard, *La fabrique du regard. Images de science et appareils de vision (XVe-XXe siècle)* (Paris: Odile Jacob, 1998), 182.

talented philosophers? Consciously or subconsciously, humans like it when a spark of heart and soul slips in between the too tidy, too shiny pages of catalogues of satellite views of our planet. Even if a bit of objectivity must consequently be sacrificed.

A second ethical question regarding the views of Earth from space could involve the apparent fragility of our planet, which adds a dramatic aspect to the aesthetic experience of the overview effect. How can we explain this impression of fragility? Thanks to astronauts and satellite images, humanity can enjoy the extraordinary view of the diversity of natural and artificial spaces composing the surface of the planet today. In addition, this spectacle appears as a dynamic process. The Swiss astrophysicist Claude Nicollier, who was an astronaut for the ESA in four missions onboard an American space shuttle, speaks of the 'endless succession', of the 'constant renewal of the overflown landscape'. He recounts that

> From Morocco to the Red Sea, you cross the Sahara desert in less than ten minutes. Ten minutes later, you're over the Himalayas. Splendid snowy mountain caps, interspersed with glaciers with, in the South, the foggy plains of the Ganges and Brahmaputra rivers and, in the North, the high Tibetan plateau, peppered with uncountable bright blue lakes.[4]

Even if it is artificial, does this dynamic, this acceleration not contribute to creating the impression of fragility that astronauts mention so frequently? Especially when satellites add to this their ability to pass regularly above a same area and thereby highlight the evolving, dynamic nature of the earthly environment: alternation between day and night, succession of seasons, changes in geographic, biologic, or hydrologic conditions, etc. This dynamic aspect, however, is perhaps not as easy to accept as we would like to think. In the minds of many of our contemporaries, nature is still perceived as a sort of cosmos, in other words as a reality that has been set once and for all, or at least as something that only presents known and limited variations. Every evolution, every noticeable change is seen as an offense to the majestic and (supposedly) inalterable beauty of the world, which we prefer to speak of in terms

4.   Claude Nicollier, 'Une majestueuse beauté (A majestic beauty)», in *Les Conquêtes de l'espace. Raisons et passions d'un défi*, *Monde diplomatique – Savoirs*, n°3 (1994): 22.

of conservation or preservation, or possibly restoration or renovation. The fact of noticing changes, sometimes even disruptions in the heart of nature can, therefore, take on a dramatic aspect, especially if the observation in question puts us face to face with our responsibilities.

Here again, the objectivity of photos is insufficient and words must accompany them in order to add, not a spark of humanity in this case, but a supplement of explanation and awareness. Let's take the example of the Amazon forest: photos taken from space, the first of which date back to the late 1980s, depict, better than any map or speech, what specialists call the 'gnawing' of the Amazonian forest cover (the French expression for this is 'mitage', referring to the impression of a piece of fabric attacked by moths). Nonetheless, these photos don't only serve to assess the rate of deforestation, which is estimated today to be around twenty per cent of the total Brazilian forest cover during the past forty years, that is the equivalent of one football field every seven seconds. Indeed, observation from space also leads us to raise qualitative questions: what processes for extracting lumber are being employed? Via which road network is this lumber being transported? How can one explain the phenomenon of 'greening' of deforested areas, which can be observed from one photo to the next? Is it a planted crop, grass destined to livestock, or simply a natural regeneration of the forest? 'Put into photos', deforestation is no longer merely a matter of numbers. Its biological and ecological issues (the management of biodiversity and of genetic resources), as well as its social, economic, and political issues (issues of land ownership and agricultural implantation, for example) become easier to identify.

It seems like a very long time since Antoine de Saint-Exupéry, witnessing the overview effect from his airplane, could write in *Southern Mail* about

> a well-ordered world beneath him—3,000 meters—neatly laid out like a toy sheepfold in its box. Houses, canals, roads—men's playthings. A sectioned world, where each field touches its fence, each park its wall. Carcassonne, where each milliner relives the life of her grandmother. Humble lives happily herded together, men's playthings neatly drawn up in their showcase. Yes, a showcase world, too exposed, too spread out, with towns laid out in order on the unrolled map and which a slow earth pulls towards him with the sureness of a tide.[5]

---

5.   Antoine de Saint-Exupéry, *Courrier Sud* (Paris: Gallimard, 1929), 29.

This orderly, cosmic image, so well described by the French pilot-writer, has now been replaced by the dynamic image of a world submitted to complete change—the image of an apparent and worrying fragility.

We must heed the warning of Jeffrey Hoffman, who travelled onboard a space shuttle on five occasions: 'It would be naive to suppose that all we have to do is take some photos of the Earth from space and promote an ecological awareness for all problems to be resolved'.[6] In truth, the shock inspired by photos and the weight of words do not suffice to trigger the feeling of fear that Hans Jonas associates with the principle of responsibility, with an awareness that would match the needs of the situation. Of course, taking the full measure of the situation of our planet does not fall to space technologies alone. Nevertheless, space is capable of providing the means and data to assess this situation and to realise that our perception of the Earth is partially constructed by space itself.

## Is the Earth a spaceship?

Since the end of the 1950s, the space enterprise has made us aware of our earthly condition by offering us a global view of our planet within the universe. On February 14, 1990, NASA gave Voyager 1 the order to do a series of portraits of the various planets it visited since it left the Earth on September 5, 1977. Our planet, then located 6,4 billion kilometers away, appeared as a 'Pale Blue Dot', as Carl Sagan called it. He added 'Look at this little dot. Here it is. That's our home. That's us.' And he concluded 'There is perhaps no better illustration of the folly of Man's ideas than this distant image of our tiny world. For me, it underlines our responsibility to cohabit with one another in a brotherly way, and to preserve and cherish the pale blue dot, the only home we've ever had.'[7] Thus, the Earth is not only our cradle, which we could eventually leave some day, if the necessity arose. It is still—and will remain for a long time—our homeland, to which we return after our interplanetary races, and which our gazes, directed towards

---

6.  Jeffrey A Hoffman, 'A personal account of spaceflight', in Jean Schneider and Monique Léger-Orine, editors, *Frontières et conquête spatiale. La philosophie à l'épreuve* (Dordrecht/Boston/London: Kluwer Academic Publishers, 1988), 203.
7.  *Cf* Carl Sagan, *Pale Blue Dot: A Vision of the Human Future in Space* (New York: Random House, 1994).

the cosmic emptiness, always end up returning to. Through the space enterprise, the Earth has gone back to being the prime destination of the human odyssey. Therefore, it is not surprising that spacefaring men and women like to refer to the Earth as a spaceship.

The idea is appealing, especially when it is suggested by an astronaut who can testify of both the planet's apparent fragility and the care required by an artificial structure as complex as a space station, in order for it to ensure a suitable shelter for its human occupants at the heart of an environment as hostile as outer space. Henceforth, doesn't it seem obvious, explicit, and educational to compare the Earth to a spaceship and the inhabitants of the Earth to its crew? Yet once more, it makes sense to question the relevance of such evident semantics.

Apparently, we owe this idea to Richard Buckminster Fuller. This futurist American architect, renowned for his geodesic domes, is said to have used the metaphor of 'Spaceship Earth' as early as the 1950s. It became really famous, however, with the burgeoning of the space era and then with the publication, in 1969, of one of his main works, 'Operating Manual for Spaceship Earth'. In that essay, Fuller described humanity's situation and the challenges it was facing. He advocated a global vision and a systematic approach, and promoted a spirit of innovation and local action. To ensure the viability or merely the survival of the human species on a planet with limited resources—many of which are non-renewable—it is imperative to 'do more with less'. Fuller's words are not really alarmist. On the contrary, they were uttered by an architect who was convinced that it would be possible to gain great enough knowledge of the Earth's vital and regenerative systems to master its functioning and avoid a major crisis. In other words, to turn our planet into a technical artifact, a machine. Fuller designed Spaceship Earth like the geodesic domes which he invented, or the ships on which he sailed when he was in the army. A technocratic vision and a general mobilisation seemed essential to him to steer Spaceship Earth and keep it 'afloat'. When the *Manual* was published, and later too, when some of Fuller's ideas were taken up by James Lovelock with his Gaia hypothesis (this English scientist considered the Earth as a single living organism), there was no lack of criticism to denounce the possible dangers and risks of a planetary technocracy, a suppression of democracy in favor of a totalitarian power. Indeed, according to Fuller, the master of the vessel (who could be none other than the engineer-architect) must impose strict control when confronted with potential restrictions or shortages of natural resources, as well as an increase in population.

The metaphor of the terrestrial spaceship leads us to examine critically the notions of geo-engineering and the idea that tomorrow's technologies could solve all disorders, excesses, and catastrophes caused by yesterday's and today's technologies. It also makes us raise questions pertaining to the humanities and moral sciences by introducing a concept that belongs to the realm of science-fiction; that is generational vessels. The principle is simple. Only vessels capable of hosting dozens of successive generations of explorers or fugitives have a chance of reaching other stellar systems or galaxies. Those who imagined them (Robert H Goddard, Konstantin Tsolkiovsky, John D Bernal) speak either of generational vessels, colony-vessels, or spatial arks.

It is fascinating to imagine a group of humans living onboard a vessel during several generations and in total autonomy. They must find ways to ensure not only their respiration, nourishment, protection from cosmic rays, and management of gravity, but also their government, their reproduction and, ultimately, their motivation. In short, it is a matter of constructing a true biosphere, an authentic little planet Earth. This constitutes not only a scientific and technological challenge, but also a social, political, and judicial one. That is why humanities have also found interest in projects of world-vessels that share characteristics with utopias.

Will it be an easy task to select the thousands of inhabitants of these vessels, to manage their individual and collective existences, their genetic and mental health, their reproduction? It is always possible and often easy to refer, in humankind's past, to the modest size of migrating populations, but do we really know how they function? Would we accept them onboard a generational vessel? It seems necessary to set up fairly stringent rules and policies regarding birth control, therapeutic abortion, euthanasia, advanced techniques in medically assisted procreation or genetic engineering, etc. Will eugenics have to be explicitly allowed? Will it be necessary to resort to techniques that presently belong to the field of transhumanism, in order to have humans embark onboard world-vessels? These tricky questions are not the only ones . . . How to deal with the risk of losing knowledge in a self-contained society? How to prevent voyagers from forgetting the original goal of their mission? How to motivate the intermediary generations, who will not have known life on Earth and who will never reach the hypothetical destination either? There again, there is much food for thought, especially considering the current human and terrestrial affairs, since the various limits of our 'fabricated' planet, of our own spaceship, are becoming increasingly obvious and could soon be reached. Should we push the utopian exercise or the ethical questioning as far as possible, raising questions that are controversial or considered to this day as 'politically incorrect'? There is no doubt that the distinction between goal and destination, introduced by the Augustin report in 2009, is relevant—whether the objective be our Earth or the space around it, waiting to be explored.

# Chapter 6
# The Space Odyssey

Why should September 8 1966, be a date to remember in the history of space exploration? In February of that same year, the Soviets succeeded in landing the Luna 9 space probe on the surface of the Moon and, one month later, they 'impacted' planet Venus with the Venera 3 probe. In March, the Americans achieved the first spatial docking between Gemini 8 and an Agena vessel and, in August, they started mapping the Moon thanks to the Lunar Orbiter 1 probe. They were also preparing the first Apollo mission, scheduled for the beginning of the following year. So, what's special about September 8, 1966? That was when the first episode of the *Star Trek* series was screened on television, enabling viewers to discover the now famous title sequence 'Space: the final frontier. These are the voyages of the starship *Enterprise*. Its five-year mission: to explore strange new worlds, to seek out new life and new civilizations, to boldly go where no man has gone before.'

Contemporary with humankind's first steps in space, *Star Trek* played a singular role against the spatial cultural background. After Jules Verne's and Chesley Bonestell's stories and the film *2001: A Space Odyssey* (1968), the TV series became a true icon of space exploration. It inspired it and embodied it, both at the same time. The broadcasting of *Star Trek* made it impossible for anyone to remain unaware that space was, henceforth, for humankind, the final frontier. How are we to understand, to interpret this famous catchphrase?

## We explore as we breathe

In the *Leviathan*, Thomas Hobbes put forth the idea that 'the desire to know why, and how, [is] curiosity'. Similarly, the North Pole explorer,

Fridtjof Nansen, justified his own expeditions to the pole in the following way: 'The history of the human race is a continual struggle from darkness towards light. It is, therefore, of no purpose to discuss the use of knowledge; man wants to know, and when he ceases to do so, he is no longer man.' Curiosity (that is the will to understand why and how things and beings are as they are) would then be the main motive behind exploration, inspiring and underlying a great part of humankind's history. Should exploration hence be seen as a human specificity? Biologists, and ethologists in particular, have observed many animal behaviors that undoubtedly stem from curiosity. Introduce a strange motion, an unusual thing, or a remarkable being into their environment, and you will see dogs straighten their ears, cats open their eyes wide, and humans be startled. Their attention is sharpened. Their astonishment can either turn into fear or terror, leading to flight, or, on the contrary, turn into attraction and curiosity. When a cat plays with its shadow, or with its master's shadow, should we simply say that it is moved by a desire to understand how this phenomenon functions, or should we add that it also experiences its own incapacity to understand the phenomenon itself? The latter is likely to be true because, even if the cat cannot express it explicitly, its jumping around and desperate pawing suggests that it is experiencing an increased sense of confusion or annoyance, which isn't the case when it is following the frenetic maneuvers of one of its prey. Conversely, the 'why', the cause of the optical phenomenon, probably doesn't fit into the limits of its consciousness. As far as we know, humans are the only beings endowed with the ability to ask (themselves) such questions—questions that practically belong to the realm of metaphysics. In fact, this ability probably ensues from another one, seemingly reserved for humans too: the ability to imagine.

Imagination . . . It is because they are capable of exiting their own immediacy, of crossing the borders of time and space that human beings are obsessed by the question 'why'. When children bombard their parents with endless questions starting with 'Daddy, Mommy, tell me, why . . .?', they do so because they find themselves, thanks to their imagination, capable of taking a step back from an event or, on the contrary, capable of believing that they are actively participating in it, or of putting themselves in someone else's shoes. This experience is common to all of us when we ask 'why', or in other words, when we wonder about the origin and the purpose of life, its meaning and finality, in between the two unique experiences of birth and death. Would we do this if we didn't have any imagination?

The combination of curiosity and imagination seems to be characteristic of human beings and to constitute one of their main motivations for undertaking individual and collective explorations. Indeed, they have never explored worlds that they had not imagined beforehand, lands which they had not first dreamt about. All territories explored by humans have first seen their hypothetical borders drawn on maps and their lands populated by extraordinary beings, or left uninhabited. These are the fascinating *terrae incognitae* of the old times, the breathtaking 'final frontiers' of modern days, which do not only serve to fill the space beyond the geographic horizon, to feed the curiosity of the most adventurous humans, but also to offer glimpses of answers to the most important questions, namely those about our origins and our fate, our identity and the possible existence of others. Exploration is not a game (an 'action with no why' as philosophers like to put it), but one of the most serious and common human enterprises. We explore as we breathe.

The birth of a little human marks its entrance into the circle of explorers: it is brutally expelled, chased from the original paradise in which it had lived for nine months. It is brought into a world which it had so far perceived and, doubtlessly, imagined beyond the horizon of its mother's belly. In an instant, it discovers air, light, noises, odors, shocks, all this without the protective maternal filter. And then it starts breathing. First inspiration, first expiration. The unfolding of its pulmonary alveoli resembles the flapping of an explorer's sails amid a storm. An ultimate suffering, a harrowing cry before calm returns and breathing settles in. Until its death, the little human will never cease to inhale and exhale, to suck in and expel, to concentrate and dilate the air which it needs to live, to survive. And it is not only a matter of air: the inner and outer worlds, the self and the non-self, the known and the unknown never cease, during the course of an entire human life, to meet, exchange, and confront each other on the thresholds of our senses, our knowledge, and our consciousness. Yes, Man explores as he breathes, and he breathes as he explores.

Exploration is and will always remain a dramatic enterprise that cannot avoid cries, screams, and tears, both of joy and pleasure, and of sadness and suffering. Exploring is a human endeavor, so terribly and magnificently human that it cannot be met by any other end than death. *Usque ad mortem*, until death, for that is the ultimate ending of human exploration. Thus, for humankind, exploration seems to be more than self-obvious.

## Who goes there?

The short-story entitled *The Sentinel*, written by Arthur C Clarke, and even more so, *2001: A Space Odyssey*, the film directed by Stanley Kubrick based on this short-story, both depict one of the main dynamics of human history and of its many exploratory episodes: the search, the quest for the other. From the primate-parents' confrontation, hitting each other with stones and bones, to the enigmatic return of the hero in the form of a fetus, and to his clash with the cold stubbornness of computer HAL, all this under the influence of a mysterious extraterrestrial monolith, Kubrick and Clarke tell us of nothing else than the fascination of the human mind for the possible existence of someone or something other than itself. Regardless of its identity—it may be our neighbor living in the next cave or our *alter ego* built with electronic chips, a powerful enemy on the other side of the border or a hypothetical inhabitant of a distant galaxy—the other always follows us like a shadow. But in the end, when we have reached the term of our quest, the ultimate unknown, the last *alter ego* that each of us must discover can only be found within. 'I is another' wrote the French poet Arthur Rimbaud to his friend Paul Demeney, in a letter dated May 15, 1871. May the reader be reassured: space ethics do not require us to delve any deeper into psychological introspection and philosophical questioning about ourselves! However, as mentioned previously, explorers cannot escape the question of their own identity, nor of their origins. Confronting the unknown can probably help us build ourselves, our national pride, but it will also shake our references and test our common identity.

Let's point out that human cultures did not wait for the exploration of our planet to be complete before they started imagining beings even more extraordinary than the potential inhabitants of *terrae incognitae*, distant lands still to be discovered. Early on, angels and demons have populated the underworld and the heavens above. The first were meant to feed the chthonian and infernal fires, the latter to swirl around in the celestial ether, ensuring the choral animation of paradise. Many divine figures stem from an analogous psychological, mythological, or cultural process. If I add the fact that the term 'extraterrestrial' was long used as a synonym for supernatural, it will become clear to the reader that I consider the current figures of *aliens*, Martians, and other extraterrestrials as modern avatars of these 'others', who, I repeat, have accompanied the human species like its shadow since its beginnings.

At present, astrobiology is the field of scientific research that brings together and concentrates most of the efforts undertaken to answer the question of the others. Is it a science with no subject? Whoever denies this is convinced that life exists not only elsewhere, but everywhere. 'We are not on the search for the Holy Graal', claims Michel Dobrijevic, who shares the belief that extraterrestrial life is abundant. Among the supporters of this idea, many physicists apply probability calculations to the astounding number of galaxies that our universe (alone) seems to have, and to the just as astounding number of stars and planets contained in each of these galaxies. In contrast, their detractors believe that Earth alone was the stage of the emergence of biological structures as complex as those populating our planet, whose appearance required a combination of very specific conditions. These scientists often belong to the field of biology. In other words, the debate rages between those who don't want scientists to lock themselves into an excessively anthropocentric or geocentric conception of life (that is meaning counting only the Earth or the human being as valid references), and those for whom life necessarily requires biochemical criteria similar to those found on Earth. Charles Cockell notes that 'the exploration of the planet, first *de visu*, then *in situ*, has nearly always shown that we tend to be too quick at loosening the reins of our imagination. We have had to back-track many times and return to far less ambitious hopes regarding present or past Martian life. Nonetheless, on Earth, the diversity of life is extraordinary'. According to the British Antarctic Survey researcher, zoos, whether terrestrial or extraterrestrial, have fences and limits to the scope of their possibilities. He concludes 'In my opinion, our earthly zoo rather represents the norm. Hence, basing ourselves on the criteria that define its perimeter does not seem to be an unacceptable error factor in the quest for life in the universe'[1].

---

1. Cockell, Charles. 'Les grilles du zoo extraterrestre (The fences of the extraterrestrial zoo)'. *La Recherche*, March 2001: 33–34.

Indeed, defining the perimeter of the living is the first challenge astrobiologists must face. This issue is not purely scientific. It quickly takes on a philosophical turn, since it concerns the very notion of life. In his *Confessions,* Augustin d'Hippone wrote about time along similar lines: 'What, then, is time? If no one asks me, I know what it is. If I wish to explain it to him who asks me, I do not know.' Today, a living being could be defined as an entity capable of duplicating itself and evolving, thanks to metabolic mechanisms and the use of available sources of energy (heat or solar radiation, for example). Viruses are already on the margin of the living, because they require the DNA of other micro-organisms to duplicate themselves. If we look at the prion, which is responsible for bovine spongiform encephalopathy (BSE), the situation gets even trickier: it is a pathogenic entity, transmissible but not alive. A prion is no more than a singular protein that is extremely difficult to destroy with current processes of sterilization. Henceforth, researching and analysing forms of life, or identifying reliable biological markers (that is molecules or organic components characteristic of the living) can be seen as a true challenge for biologists. At the same time, and over the last few decades, the discovery of extremophiles (that is organisms living in conditions so far deemed 'uninhabitable') did not only surprise scientists. It also led to them to entirely reconsider the phylogenetic tree of living organisms on Earth. Thus, the revolution brought about by astrobiologists broadens even more the issue of life and its origins. The classic sentinel's question 'Who goes there?' can now be applied to hypothetical extraterrestrial organisms as well as to terrestrial ones.

If the question of the other's identity can generally remain purely theoretical and be set exclusively within the limits of knowledge, it becomes practical when a possible encounter has to be envisaged—a 'close encounter of the third kind', according to Hynek's classification.[2] Without having to picture any sort of UFO, extraterrestrial vessel, or flying saucer, the question arises today among scientists who deal with the expedition of earthly machines or humans to another celestial body, or with the return trip of said machines, humans, or extraterrestrial samples. Whether the concern is to reduce risks of contamination

---

2.  Hynek's classification is a method to classify UFO sightings. It was proposed by the American astronomer J Allen Hynek, in 1972, in his book *The UFO Experience: A Scientific Study.* This classification lies at the origin of the title *Encounter of the third kind,* a film directed by Steven Spielberg in 1977, showing close encounters going crescendo from 1 to 5.

of the Earth and its living organisms, of astronauts, or of the explored planets, what scientists and engineers describe today as 'planetary protection' probably constitutes one of the main challenges of present space exploration. Indeed, it is directly and precisely involved with a field of human knowledge that remains unexplored, and for which the question 'Who goes there?' has so far received very few answers.

---

## COSPAR's Planetary Protection Policy

COSPAR (Committee on Space Research) is an international group in charge of organizing scientific projects associated with space exploration. It was created in 1958 by the International Science Council (ISC).

COSPAR has progressively established rules of planetary protection which depend on the nature of the mission and on the celestial body aimed for. The risk of contamination and the need for protection are lowest in the case of a simple overflight of a planetary body (in which case the risk is linked to a possible navigation error, leading to a crash of the space probe). The risk increases when it is a matter of orbiting a planet or landing on a fixed station. It reaches its maximum levels when one envisages landing on the planet or moving around on its surface. The risk of contamination also varies according to which celestial body is aimed for, as its properties can be deemed more or less favorable to the development of life.

Consequently, COSPAR has defined five categories of missions:

* Category I includes any mission to a target body which is not of direct interest for understanding the process of chemical evolution or the origin of life. This involves undifferentiated, metamorphosed asteroids; Io. No protection of such bodies is warranted and no planetary protection requirements are imposed by this policy.

* Category II missions comprise all types of missions to those target bodies where there is significant interest relative to the process of chemical evolution and the origin of life, but where there is only a remote chance that contamination carried by a spacecraft could compromise future investigations. This involves Venus; Moon (with organic inventory); Comets; Carbonaceous Chondrite Asteroids; Jupiter; Saturn; Uranus; Neptune; Ganymede; Callisto;

Titan; Triton; Pluto/Charon; Ceres; Kuiper-Belt Objects > 1/2 the size of Pluto; Kuiper-Belt Objects < 1/2 the size of Pluto. The requirements are for simple documentation only. Preparation of a short planetary protection plan is required for these flight projects primarily to outline intended or potential impact targets, brief Pre—and Post-launch analyses detailing impact strategies, and a Post-encounter and End-of-Mission Report which will provide the location of impact if such an event occurs.

* Category III missions comprise certain types of missions (mostly flyby and orbiter) to a target body of chemical evolution and/or origin of life interest and for which scientific opinion provides a significant chance of contamination which could compromise future investigations. This involves orbiters of Mars, Europa, Enceladus. Requirements will consist of documentation (more involved than Category II) and some implementing procedures, including trajectory biasing, the use of cleanrooms during spacecraft assembly and testing, and possibly bioburden reduction. Although no impact is intended for Category III missions, an inventory of bulk constituent organics is required if the probability of impact is significant.

* Category IV missions comprise certain types of missions (mostly probe and lander) to a target body of chemical evolution and/or origin of life interest and for which scientific opinion provides a significant chance of contamination which could compromise future investigations. This involves lander missions to Mars, Europa, Enceladus. Requirements imposed include rather detailed documentation (more involved than Category III), including a bioassay to enumerate the bioburden, a probability of contamination analysis, an inventory of the bulk constituent organics and an increased number of implementing procedures. The implementing procedures required may include trajectory biasing, cleanrooms, bioburden reduction, possible partial sterilization of the direct contact hardware and a bioshield for that hardware. Generally, the requirements and compliance are similar to Viking, with the exception of complete lander/probe sterilization.

* Category V missions comprise all Earth-return missions. The concern for these missions is the protection of the terrestrial system, the Earth, and the Moon. (The Moon must be protected from back contamination to retain freedom from planetary protection

requirements on Earth-Moon travel.) For solar system bodies deemed by scientific opinion to have no indigenous life forms, a subcategory "unrestricted Earth return" is defined. Missions in this subcategory have planetary protection requirements on the outbound phase only, corresponding to the category of that phase (typically Category I or II). For all other Category V missions, in a subcategory defined as 'restricted Earth return', the highest degree of concern is expressed by the absolute prohibition of destructive impact upon return, the need for containment throughout the return phase of all returned hardware which directly contacted the target body or unsterilised material from the body, and the need for containement of any unsterilised sample collected and returned to Earth. Post-mission, there is a need to conduct timely analyses of any unsterilisd sample collected and returned to Earth, under strict containment, and using the most sensitive techniques. If any sign of the existence of a nonterrestrial replicating entity is found, the returned sample must remain contained unless treated by an effective sterilising procedure. Category V concerns are reflected in requirements that encompass those of Category IV plus a continuing monitoring of project activities, studies and research (that is, in sterilization procedures and containment techniques).

COSPAR has added sub-categories for IV and V, according to the instruments used (search for life or not), the sites visited, and the planets from which the samples are taken.

Humankind has forever lost its innocence. More than thirty years after the Tchernobyl disaster and twenty-five years after the Rio de Janeiro conference, signs of global warming are multiplying. Natural environment degradation and resource depletion are worsening. Humans are becoming aware of the fact that they are dramatically disturbing the Earth's fragile balance, even though this balance may seem only momentary on the scale of cosmic time. Many of our contemporaries wonder whether we will be able to make it through this 'century of threats' (Jacques Blamont), which began with the new millennium. When they found out that this terrible ability to disturb the environment had been exported all the way into the circumterrestrial space, and then even into outer space, these same people had the right to raise a few questions.

What are our precautionary measures actually worth, especially with regard to political and economic stakes? What should we expect from space law and principles established half a century ago? What sort of responsibility can we honestly and reasonably have towards planets that are still unknown or imperfectly known? Can a planet, susceptible of hosting forms of life or precursors thereof, be protected from all terrestrial biological intrusion? Should it be? These questions are all the more pressing as 'harm' risks being done sooner than later and, most likely, in an irreversible way. Not to mention the risk we are running of contaminating our own planet in return. For all these reasons, it is not surprising that the media and public opinion are so critical of space actors and their projects and commitments, when the question of space pollution is broached (see next chapter: 'Has space become a trashcan?'). It is not surprising either that these same space actors avoid using terms such as 'contamination' or 'pollution'. Spreading information about a state that escapes normality is a delicate art that requires finding balance between triggering panic and numbing vigilance. Ethical questioning can have a role to play here.

## Risk is part of the future

To the question 'Who goes there?' we can add another one that has subsisted throughout the modern history of science and technology, and has always been part of humankind's interrogations: what right do we have to modify Nature, and to what extent? In other words, may we voluntarily and artificially create an alterity? This question can be formulated differently: are there limits to human activities, depending on whether they belong to a natural process or whether, on the contrary, they transgress it? Public reactions will depend on the various answers suggested by scientists. Some scientists hypothesise that, since the solar system was formed some 4.5 billion years ago, biological cross-contamination has been a common and frequent phenomenon on its various bodies. Falling comets filled with water and organic chemical compounds are 'responsible' for this, as well as asteroids or meteorites «ejected» from other planets. Thus, when humans transport Earth-dwelling organisms to another planet, are they not merely accomplishing a natural process? Once the issue of planetary protection has been solved within the scientific framework of the study of extraterrestrial organisms, this question deserves to be raised. It illustrates how worthy the opinion of one of my colleagues at the CNES is, on the subject of ethics: 'Taking ethical questioning seriously does not prevent us from working today, but it makes us commit to the future.' The desire to know and the effort to explore are then met by the obligation to choose, which requires us to be reasonably aware of the existing risks.

My point is not to dwell on a history or a philosophy of risk, but rather to describe the main elements that explain how modern societies understand risk, based on how they are influenced by scientific knowledge and technical progress.

Let's first note how, to discover the risk, or rather to elaborate a modern version of it and to fully experience it, humans have not only had to change their world, but also their conception of the world and, in particular, reject the idea of the cosmos that I described earlier ('an orderly, harmonious system'), with its philosophical foundations and religious convictions. To reach this point, explorers have had to be brave enough to cross the last terrestrial and maritime borders, astronomers have had to work hard at inventing an unlimited universe instead of a cosmos closed in on itself, and philosophers have had to announce the death of God and face the drama experienced by humanity, once it had lost the assurance provided by its belief in an all-mighty providence. Chance had to be summoned, although not in the old ways through fun or divinatory methods, but with the aim of describing it, calculating it, and trying to control it. This task was first entrusted to mathematicians. At the end of the seventeenth century, they laid down the foundations for the concept of probability and defined mathematical expectation as a sort of center of gravity among all possible consequences of a similar process or event, whether successful or unsuccessful, beneficial or harmful. There was no longer any talk of relying on cosmic assurance or fate. The human person had to learn to decide for himself, based on an estimated value of his acts and their consequences. He had to become a calculator and, at the same time, take responsibility for his choices as well as for his actions. Henceforth, he would have to be stronger than the gods.

The history of modern astronautics is full of programs dead in the water, of aborted launching procedures, of rockets exploding on their launching pads or a few kilometers into the air, of lost space probes, and of damaged capsules and stations. That is why engineers now systematically calculate the risks to the machines they build and to those who use them or travel in them. They have never really believed in what is commonly called 'zero risk', and which actually corresponds to zero failure. Alongside French philosopher François Ewald, they recognise that the human is 'an animal destined to risk' and that, in many situations, he can no longer impute situations of threat or failure to external causes. Ulrich Beck explains 'The sources of danger are no longer ignorance but knowledge; not a deficient but a perfected mastery over nature'.[3] Thus, and more generally, when defined as the combined effect of chance and necessity, risk constitutes one of the central elements of the modern human condition. Nonetheless, it doesn't provide any meaning, any end in itself. Risk is a more or less prospective thing, that becomes outdated as time goes by. It never actually belongs to the past, like a nostalgic or satisfied sigh. On the contrary, it is always part of the future and can thereby never be dissociated from exploration.

Questioning an exploration project while introducing the risk factor implies raising the question of its opportuneness. How opportune is it, time-wise? Is now the best time or would it be better to postpone the departure, to establish a deadline, or request a moratorium? Similarly, how appropriate are the means involved? So many promises have been made to expand frontiers, discover new worlds or meet new people, promises that were repeated and defended, but without their authors having sufficient financial, human, and technical means to keep them. Before ships even weigh anchor, before caravans leave the last known post, or before rockets are launched from their pad, the most lucid minds must take the time and invest the necessary means to properly assess the chances of success of any given mission. The world of astronautics mustn't forget the tragic human losses that resulted from the accident of the Challenger shuttle, or the less dramatic accident of the Mars Climate Orbiter probe. In 1999, this probe missed its launch into orbit around Mars and went up into flames in the Martian atmosphere because the teams of Lockheed and NASA hadn't used the same units of measure. And lastly, what is there to say of a mission's moral opportuneness? We must learn to resist the engineer's urge to undertake something for the mere reason that it seems feasible, or that we have the *ad hoc* knowledge to do so. The famous

---

3. Ulrich Beck, *La société du risque. Sur la voie d'une nouvelle modernité* (Paris: Flammarion, 2008), 399.

expression "We can do it" has other limits beyond the acquired knowledge or the available power. Here, I use the term 'moral opportuneness' for we should consider things according to David Hume's law, known as the 'is-ought problem': everything that is possible is not necessarily desirable for the mere reason that it is feasible, even if scientists and engineers guarantee its feasibility. Space exploration has often depended and still often depends on arbitrary government acts (*fait du prince*); conversely, it should not exclusively be decided by scientists and engineers when it relies on public funds and potentially endangers lives, especially lives of non-professionals, sometimes even of entire populations. If these guidelines are respected, exploration can remain the pride of humanity as a whole.

# Chapter 7
# Questions About Space

I have now thoroughly explained and abundantly illustrated the fact that ethics are above all a matter of attitude, an interrogative practice. Why? How? With what consequences? It is not only useful but indispensable to ask these questions in order to grant our actions a truly human component, to integrate them into the societies of their time, and to undertake them with a deep concern for responsibility and sustainability.

The previous pages were aimed at suggesting ways to broach such questions and providing responses thereto. In this chapter, I would like to summarise my point and complete it by tackling four questions that I am frequently asked.

## Has space become a trashcan?

This question is very often asked to space-related workers and, above all, to space agencies. If we study the media and the topics they prefer to address (excluding specific punctual events), it even seems to be their favorite. Thus, this question must be dealt with seriously and, therefore, requires two prior clarifications.

### It is human to pollute

We commonly speak of debris, trash, waste, and nuisance, or of soiling, polluting, contaminating, and fouling. Yet the terms we use don't really matter. What counts is the observation we can make from the outset: polluting is a human and natural behavior. Indeed, waste, debris, and garbage are the most immediate and often inevitable traces not only of human activity, but of the presence of living organisms in general. At a time and in a culture where health, cleanliness, protection, and conservation are essential concerns, we often tend to forget an obvious fact: every being, from the moment it starts breathing, walking, eating, sleeping, and reproducing itself, produces residues resulting from the mere fact that it is alive. Most of the time, who or whatever generates these residues neither knows nor cares how they are managed, especially because the processes enabling their salvaging, transformation, and reuse for new purposes are numerous and generally very efficient. Without them, waste would quickly and inexorably encumber every possible biological or ecological niche. Since humans are part of nature, they fall within the scope of natural cycles, as do some of their activities. For instance, the fouled air expired by their lungs joins the atmosphere, their dirty waters flow into rivers and get filtered by groundwater tables, their feces enrich the soil's humus, as do their cadavers, regardless of which funeral rites they observe.

This being said, the human person is not only a creature of nature. He is also a being of culture and, since he first appeared on Earth, he developed activities that are foreign to nature and its processes. Hunters and gatherers have become farmers who use inputs and produce waste unknown up to then. Walkers have become drivers, aviators, and seafarers, who have broadened the areas of human occupation, transporting to all sorts of places not only the essentials and non-essentials for their survival, but also other species, either domesticated ones or parasites. Cave-dwellers built houses and buildings that can compromise the balance of any given environment and the smooth running of its cycles. Thus, to his natural waste products human beings long ago started adding those of technical and cultural activities, meaning those that depend on human consciousness and reason, on a human being's ability to foresee and to decide. Due to their very origin, these residues do not necessarily fit into natural cycles of transformation and elimination. In fact, they rarely do. On the contrary, they can obstruct them, deeply disturb them, or even break them. In such cases, they really become forms of nuisance or pollution. The human person can neither neglect them, nor trust nature alone to take care of them. Human beings must 'culturally' deal with the waste created by their own culture. Otherwise, the human being runs the risk of throwing off balance and threatening the natural and artificial environments to which he or she belongs and on which his or her existence depends.

To summarise, polluting might be human, but the fact of contemplating cultural (and sometimes natural) types of pollution, of understanding, evaluating, limiting, and isolating them, or even avoiding to produce them in the first place, contributes to the way humans see themselves as rational and free beings, members of a group, a society, a species—in short, as occupants of the remarkable planet Earth.

### Trashcans . . . an improvement!

Sick of the extreme filth of Parisian streets, in 1884, the prefect Eugène Poubelle signed a decree concerning the disposal of domestic garbage. He imposed on all building owners to provide their tenants with communal containers destined to receive household waste. He seized this opportunity to invent selective sorting: one container was meant for 'household refuse', another for glass and ceramics, and a third one for oyster shells. Parisians, who had up to then practiced the method of "everything onto the street", were not fond of their prefect's initiative. To take revenge for it, they gave his name to these containers that were equipped with a lid and a handle. The success of this naming was immediate and, as early as 1890, the word *poubelle* entered the supplement of the *Grand dictionnaire universel du XIXe siècle* (Great Universal Dictionary of the 19th Century).

So, in the strict sense of the term, space is not a 'trashcan'. It doesn't even dispose of such a sophisticated system of waste management. It would be more appropriate to speak of a garbage dump or a rubbish chute, even though, on Earth, such things are located in a defined perimeter, whereas spatial debris can be found anywhere in outer space, in the general area around the Earth.

---

### Debris around the Earth

COPUOS gives the following definition of spatial debris: 'All human-made objects, including fragments and elements thereof, in Earth's orbit or re-entering the atmosphere, that are nonfunctional' (*Space Debris Mitigation Guidelines of the Committee on the Peaceful Uses of Outer Space*).

The debris density in space varies considerably according to the altitude. It is highest around 850, 1,000 or 1,500 km, with an average of one object every 100 million $km^3$. Above 1,500 km, the density decreases with the altitude, except in the vicinity of the semi-synchronous orbit (20,000 km altitude) and the geosynchronous orbit (36,000 km), where it locally reaches higher levels. Other orbits, today considered devoid of interest and relatively debris-free, could some day see their status change. Hence, those situated around 2,500 km altitude seem to be of interest to the military for the installation of observation satellites. It would probably be best to start showing concern for their protection now, because once the polluting process is initiated, it becomes difficult to control it and impossible to reverse it completely.

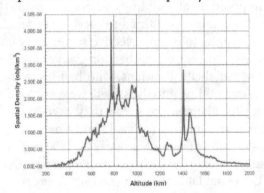

Density of space debris depending on the altitude (source: NASA) (translate x0E-08 by x.10-8)

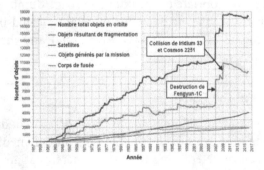

y = Number of objects

x = Year

Total number of objects in orbit

Objects resulting from fragmentation

Satellites

Objects generated by the mission

Rocket body

- Collision of Iridium 33 and Cosmos 2251
- Destruction of Fengyun-1C

Evolution, since the beginning of the space era, of the number of objects in orbit (> 10 cm in low orbit and > 1 meter in other orbits) followed by the American surveillance network USSTRATCOM. These 17,000 objects, for which we have the orbital characteristics, only represent a small fraction of the 500,000 objects measuring > 1 cm. (Situation at the end of the first day of the first trimester 2016)

The future evolution depends primarily on the number of launches, especially on their reduction recorded in Russia over the past few years. One also has to take into account the evolution of the types of missions. Constellation projects, involving dozens or even hundreds of satellites, could lead to fears of a rapid increase in the number of objects in circumterrestrial orbit, but these projects are now dormant. However, another potential risk is looming with the development of «small» satellites (mini-satellites weighing between 100 and 500 kg, microsatellites between 10 and 100 kg, nanosatellites between 1 and 10 kg, and picosatellites of less than 1 kg) that can be launched in clusters with the same launcher.

Managing space debris thus consists of various measures to prevent, detect, protect, and deorbit.

## For a sustainable space

With the United Nations Conference on Environment and Development (UNCED), also known as the 1992 Rio de Janeiro Earth Summit, the concept of 'sustainability' became common in the context of longlasting human activities. It is therefore not surprising that this concept ended up appearing in the space sector. Karl Doetsch was presiding the COPUOS when he introduced the term into the activities of the Vienna commission in 2004. He suggested to reflect upon 'the long-term sustainability of space activities' (LTSSA). His successor at the COPUOS chairmanship, Gérard Brachet, kept this idea and made it a subject of reflection for an *ad hoc* working group. The LTSSA Working group was created and its agenda adopted in 2008. Among its key activities figured the issue of space debris[1].

Let us reconsider the definition of sustainable development that the Brutland report came up with in 1987 and that inspired the Rio Summit: 'Development that meets the needs of the present without compromising the ability of future generations to meet their own needs'. This definition was adopted, repeated, and modified over the following years. Economic, sociopolitical, and eventually cultural aspects were added to the environmental dimension. The specificity of the notion of sustainable development is that it must be set into a timeframe that is neither fixist nor conservative (that is nothing must change), meaning it must envisage development as a possibility, or in other words, accept change. Nonetheless, this change may not take on just any form. Development must be carried out with respect for future generations, with a sense of intergenerational responsibility. This notion was first introduced by Hans Jonas in 1979, in his book *The Imperative of Responsibility* (*cf* Chapter 2: 'What are ethics?'): human actions may not jeopardise the possibility of future life on Earth. Thus, Hans Jonas associates the imperative of responsibility with anthropocentrism, or even biocentrism, meaning he gives priority to human beings or life in general. From an ethical point of view, Jonas' primary interest lies in the consequences of human choices and actions. In his book, he does not focus on the question 'why'.

Yet that is, in truth, the core of the question 'Has space become a trashcan?'. We may not content ourselves with imagining technical solutions to produce less space debris, to protect our satellites and spacecrafts from them, and to eliminate or collect them. We must also tirelessly wonder why we produce them. That is how, in a certain way, debris are one of the main motives justifying the introduction of ethics into space-related activities.

---

1. The four focus areas of the LTSSA working group were: A – Sustainable space utilization supporting sustainable development on Earth; B – Space debris, space operations, and tools to support space situational awareness sharing; C – Space weather; D – Regulatory regimes and guidance for new actors in the space arena.

## Is space Big Brother's ally?

'THE STRUGGLE WAS FINISHED.
HE HAD WON THE VICTORY OVER HIMSELF.
HE LOVED BIG BROTHER[2].'

The last words of *1984*, George Orwell's famous novel, sound like a statement of failure or even defeat. A power, founded on total surveillance, not only got the better of a man thirsting for freedom and independence, but it even managed to make him love his condition, his subjection. Seventy years after the publication of *1984*, we cannot help but wonder whether we haven't set up social organizations similar to those described by Orwell, societies that ensure the infamous notion of 'Big Brother is watching you!' thanks to flotillas of spy-satellites, satellites of observation and tapping, of communication and localisation.

### Human show

In 1791, British philosopher Jeremy Bentham published a text, the title of which is already a summary or, even better, a program:

> Panopticon; or, the inspection-house: containing the idea of a new principle of construction applicable to any sort of establishment, in which persons of any description are to be kept under inspection; and in particular to penitentiary-houses, prisons, houses of industry, work-houses, poor-houses, lazarettos, manufactories, hospitals, mad-houses, and schools: with a plan of management adapted to the principle.

As we know, Bentham's interest in the penitentiary sector was an opportunistic one. First and foremost, he sought to find a solution to overpopulation and the insalubrity of British prisons at his time. His architectural project rested on a principle of self-discipline, namely making the prisoners act as their own guards. He wasn't going to write 'God is watching you' (that is one of the foundations of religious morality) on the walls of their cells, but rather place the inmates in a situation where they would be convinced that earthly guards could surveil them at all times, yet without ever being quite sure whether

---

2. George Orwell, *1984* (Paris: Gallimard, 2002/1950), 417.

they were being watched or not. His panopticon was a ring-shaped building, divided into cells, each as wide as the building itself. A window was to let light in from the outside, and another was to look onto the inside of the edifice, in the middle of which was a surveillance tower. Due to the backlight, prisoners, convicts, madmen, but also sick or poor people, workers or school kids, all would find themselves captured by the light, thus exposed to the hypothetical view of the warder, to the invisible voyeur settled at the center of the facility. Each room, each cell would then be like a little theater in which the occupant would be the only actor, constantly forced to play the role expected from him by the administration of the panopticon or by society.

In his book *Surveiller et Punir* (*Discipline and Punish*), French philosopher Michel Foucault analyzed the trick which modern societies added to Bentham's panopticon concept. It consists in reversing the principle of solitary confinement. To ensure control, prisoners were no longer to be hidden away or deprived of light, but on the contrary, placed in broad day-light, so as to be visible by everyone at all times. Besides, modern satellites don't occupy the center of the world theater, of the planetary 'human show', but they hang around it like invisible guards, imperceptible voyeurs. From their orbital position, they play a part in establishing a true, efficient, planetary panopticon, which earthlings have more and more trouble evading. Another question then arises: who are the producers, directors, and managers of this planetary theater? Or, to use the words of the Latin poet Juvenal, *Sed quis custodiet ipsos custodes?*, that is 'Who then will keep watch on the guards?'

In *The Crystal Palace*, Dutch philosopher Peter Sloterdijk wrote 'Is observer he who perceives the other through a window of theory and succeeds in escaping counter-observation.'[3] The multiplication of means of observation and communication, whether terrestrial or spatial, has turned each of us into an observer, a potential voyeur, sometimes having abilities similar to those of the army barely a few decades ago. However, as Sloterdijk rightfully warns us, what about counterobservation? Can we claim to escape it? Obviously, the answer is no. Satellites are the most blatant illustration thereof. Their

---

3.   Peter Sloterdijk, *Le palais de cristal. À l'intérieur du capitalisme planétaire* (Paris: Maren Sell Éditeurs, 2006), 177.

orbital revolutions 'imprison' both the observers and the observed, to the extent that Sloterdijk's definition often becomes erroneous. We have now entered the era of the observed observer[4].

It is not the aim of this essay to attempt a full analysis of this situation. Nevertheless, we must at least admit that most people seem to content themselves with it (without going so far as 'loving Big Brother'). Ignorance, habit, and easiness are the main reasons for their acceptance. Some people even forget what freedom is and others actually enjoy being watched. In his study called *La grande surveillance* (*The great surveillance*), Claude-Marie Vadrot concludes that

> We are more or less voluntary prisoners of our anguishes and of those who feed them. Every day we build ourselves a more or less golden cage of computer science, weaving on demand the many threads of Ariadne that connect us to the Grand Computer, that insatiable Minotaur for whom the famous *Big Brother* is just a distant cousin, friendly and harmless.[5]

It seems pleasant to find here an allusion to the mythological figure of Ariadne, since she inspired those who had to find a name for the family of European launchers. This reference must be seen as an invitation to not minimise the importance of the role of space in the field of surveillance, or rather of self-surveillance, put in place by our fellow humans.

## A global village

At the beginning of the 1960s, Marshall McLuhan, a Canadian philosopher who felt neither enclosed nor imprisoned, started advocating the notion of a global village. In 1962, he wrote 'The human family now exists under conditions of a global village. We live in a single constricted space resonant with tribal drums. [. . .] The new electronic interdependence recreates the world in the image of a

---

4. I invite the reader to read the short-story by Friedrich Dürrenmatt: *The Assignment. Or, on the Observing of the Observer of the Observers* (1986), which describes this paradoxical situation of the observed observer, but also the desire, the need to be observed.

5. Claude-Marie Vadrot, *La grande surveillance. Caméras, ADN, portables, Internet . . .* (Paris: Seuil, 2007), 7–8.

global village'.[6] Fifty years later, the phenomenon described by McLuhan is still accelerating, to such an extent that we now suffer from an excess of information and even from a saturation of our ability to focus our attention. We are entitled to wonder whether the global village, which has indeed become a technical reality, is also a social and cultural reality, or whether there isn't some confusion between the vested interests of communication industries and the actual philosophical and sociohistorical reality of the users of these means of communication. This question can be added to all those raised today by the processes of globalization.

What we tend to call 'globalisation' is not a new phenomenon. The thirst of humanity to discover and accumulate riches, humankind's desire to dominate, to push back the limits of his or her vital space, but also humanity's fascination for risk and spirit of adventure, are all cornerstones which I have already described in the context of space exploration and exploitation. Henceforth, the emerging humanity already had globalization 'flowing through its veins', even if it did not yet have the means to achieve it, nor the slightest idea of what the world was actually like, before it progressively started conquering it. Although the development of maritime transport in the early fifteenth century marked the beginning of modern globalisation, this process only truly intensified at the end of the nineteenth century with a massive increase in trade and an internationalization of the economy. With World War II, this phenomenon took a new turn: without turning its back on its driving forces, yet relying on an intensification of its existing networks, it profited from the emergence of huge industrial groups and from the establishment of a powerful, integrated capital market. 'Integrated' is a key word for this new type of globalization. Marc Abélès, a renowned French anthropologist and ethnologist, underlines how 'the use of the global concept seems appropriate to render the level of integration and interconnection that has now been reached and that is translated by individuals' empiric perception, beyond their territorial attachments and their cultural identities, of belonging to a global world'.[7] Today, each of us is capable of drifting from one referential to another, of skipping from

---

6. Marshall Luhan, *The Gutenberg Galaxy. The Making of Typgraphic Man* (Toronto: University of Toronto Press, 1962), 36.
7. Marc Abélès, *Anthropologies de la globalisation* (Paris: Éditions Payot & Rivages, 2008), 8.

one scale to another, and of experiencing a compression of space and time, a form of immediacy and simultaneity that seems ever harder to escape. Thus, Sloterdijk can claim that 'Earthly globalization does not merely constitute one history among many others. It is [. . .] the unique temporal fraction in the lives of people who have discovered each other, *alias* 'humanity', and it deserves to bear the name 'history' or 'world history' in a philosophically pertinent sense'.[8] Because of its uniqueness, globalization merits critical, precise, and lucid attention from us. We owe it more than mere appraisal or rejection. It requires us to fully take stock of the effects of the acceleration of transportation, communication, and information flux, to such an extent that it has led to a synchronization of time, a superimposition of spaces. When absolutely every event, every situation, every being, and every group instantly becomes present to each one of us, when 'the history of the world contains the tribunal of the world',[9] where do we find the necessary distance for thought followed by action, or in other words, for the development of an ethical process? The human person, who has become the ultimate gauge for everything, finds his or herself abandoned to his or her own conscience, to his or her awareness of time and space, of good and evil, and also to his or her awareness of actions that have already been accomplished as well as those that have yet to be undertaken. Behind the decision to declare astronauts 'envoys of humankind', there is a singular and dizzying responsibility . . . a global responsibility!

## The principle of vigilance

'Yes, we call on all European governments, on the Europe of Twelve, to use all means available, without ruling out recourse to force if necessary, to end the war. Tomorrow, they won't be able to say they didn't know, they won't be able to say they couldn't do anything.' With these words, on November 21 1992, French journalist Jacques Julliard ended a silent demonstration against the policy of ethnic purification of President Milosevic's regime.[10] The same formula was used

---

8. Sloterdijk, *Le palais de cristal,* 26.
9. Sloterdijk, *Le palais de cristal,* 23.
10. *Cf* Jacques Julliard, 'Nous ne pourrons pas dire que nous ne savions pas. (We won't be able to say we didn't know.)', *Esprit,* January 1993: 138–139.

on other occasions, namely by the French President, Jacques Chirac, during the World Summit on Sustainable Development in Johannesburg, in September 2002. Or in the mid-1980s by a non-governmental organization trying to alert public opinion about the disastrous situation of the people living in the Sahel. It seems to me to be the clearest expression of what I call the principle of vigilance. Thereby, I am referring to the tension between knowledge, power, and action; between what has already been done and what can still be achieved. We will no longer be able to say we didn't know. And therefore, we run the risk of experiencing the shame described by Peter Sloterdijk, 'the shame experienced today by every waking creature, even more so than original sin; the fact of not rising up strongly enough against the omnipresent abasement of the living'.[11]

We can find many reasons to excuse, explain, or justify our inaction, our disengagement: the extent of the tasks to accomplish and the paucity of our means to do so; the existence of organizations that are more competent and better equipped to intervene; possibly the respect of individual, collective, or national freedom. 'A bit of shame is quickly forgotten' is the response with which we satisfy ourselves, provided we even feel shameful, and nothing indicates that this is always the case. Once again, we mustn't confuse the free circulation of people and goods, or the globalization of information and knowledge, with the term 'globalisation' that refers to the ability to articulate local and global scales both in thought and in action. This ability must effectively be applied, which is what ethics invite us to do. Space gives us a good example of this.

In July 1999, in Vienna, during the international conference on the peaceful uses of outer space UNISPACE III, the French and European space agencies (CNES and ESA) decided to coordinate their abilities regarding satellite data acquisition and delivery. *A priori*, the aim was not to counter the rise of American companies commercializing satellite images, but to be able to give such data cost free to any country touched by major natural or human-made disasters. Thus, they agreed upon the International Charter 'Space and major disasters' which was joined by the Canadian space agency as early as October 20, 2000 and, thereafter, by various organizations, space-related or not, from many countries: India and China, the United Kingdom and

---

11. Sloterdijk, *Le palais de cristal*, 211.

the United States, Japan, etc. In 2019, seventeen organizations were members of the Charter and it counted nineteen co-operating bodies.

This is an amazing coalition as it goes beyond the usual political and economic divides. Space agencies as well as national or international managers of space systems are likely to become members of the Charter. The organizations of civil protection, rescue, defense, and national security of any of the members of the Charter become *de facto* authorised to benefit from it.[12]

What are we to understand by 'natural or technological disaster'? It refers to any situation involving the loss of many human lives, substantial material damage on a great scale ensuing from a natural phenomenon (namely cyclones, tornadoes, earth-quakes, tsunamis, floods, forest fires, and volcanic eruptions) or from a technological accident (leading to chemical or radioactive contamination). If confronted by such events and exclusively following an explicit request on behalf of the country involved, the signatory agencies of this Charter commit to acquire and provide satellite data likely to help the populations of the affected areas. No need here to delve into the details of the procedure setting the Charter into motion. The main point is that it strives to ensure a permanent vigil (an operator is on call 24 hours a day, 7 days a week), to send images as fast as possible to the people and services which need them, to program specific photos to be taken and, before anything else, to check on the relevance and honesty of the request.

Since it was implemented in February 2002, the Charter was activated over six hundred times. The 600th activation was on March 20, 2019. This number often surprises the general public who is just as startled by the effective ability of these space agencies to pool sensitive and costly means, as by the amount of 'major disasters' striking human populations (mainly floods, hurricanes, typhoons, volcanic events, and earthquakes; more rarely oil spills and transportation accidents). No continent is spared, and many countries have already had to resort to the Charter. To cite only its first activations: February 4 2002, in the valleys of the Meuse and Moselle in Europe; on the same day, the Democratic Republic of Congo; on April 9 Afghanistan . . .

This initiative on behalf of the international space community shouldn't let us forget the shame evoked by Sloterdijk or the similar

---

12. Cf the website dedicated to the Charter: https://disasterscharter.org

denunciations by Julliard. In terms of vigilance and 'assistance to persons or populations in danger', our reason and common sense tell us that we can always do better. Nonetheless, let's consider this Charter as a lesson or an example to follow.

If Big Brother's spatial eyes can be made to serve just causes, we must nevertheless remain attentive to the conditions of this use, whether they are imposed on us or whether we agree to them. We must admit that each one of us accepts, sometimes with indifference, to be kept on file, watched, located, and traced, even without being aware of it. We quickly forget the flotillas of satellites that go round and round above our heads, as well as the countless surveillance cameras scattered throughout our streets. There is a real risk here; namely, that of confusing freedom with the dual, very contemporary process of identifying oneself with others through globalization and feeling individually protected. But freedom is threatened when surveillance is reduced to a function of security, and when the role of vigilance is restricted to guaranteeing conformity to a pre-established plan, to an imposed order and a maintained *status quo*. Is our freedom not endangered when our brother has become no more than a guard? On the contrary, freedom calls for and even demands a real sense of responsibility.

## Is space for sale?

On December 1 2012, I bought a plot of land on the Moon. The property deed has the number 369/1559 and was signed by Francis P William, lunar ambassador for the United Kingdom and Europe through delegation of authority from Dennis Hope, (self-proclaimed) president of the galactic government and head of the lunar embassy for earthlings. I paid the sum of $19,99 to purchase the area of an acre (approximately 4000 m$^2$), to which I had to add $1,51 of lunar tax and $10 for postage of the property deed. Hope claims to have sold in this fashion 600 million acres on the Moon, 300 on Mars, and 120 on Mercury, Venus and Io (i.e. respectively approximately 2.4 million hectares, 1.2 million hectares and 0.48 million hectares). He likes to name his most famous clients: George Lucas, Ron Howard, Tom Hanks, Harrison Ford, John Travolta, Meg Ryan, Clint Eastwood, Jimmy Carter and Ronald Reagan, various American and Russian astronauts, in adition to the Hilton and Marriot hotel chains.

The business of this inhabitant of Nevada, based on an efficient trade policy and a world-wide network of retailers and ambassadors, seems to be quite lucrative. But in fact, he wasn't the first to sell the Moon.

### They sold the Moon

In its edition of January 3 1808, the London newspaper *The Examiner* accused Napoleon Bonaparte of having cosmic ambitions: 'Then, I will be able to form an army of balloons, of which Garnerin will be the general, and take possession of the Comet. That will enable me to conquer the solar system, after which I will go to the other systems with my armies and, at last, I think, I shall meet the Devil.' Carried away by his literary imagination, the journalist forgot that, once back from his Egyptian campaign, in July 1799, Bonaparte ordered the closure of both the balloon factory and the training school for aerostat pilots, created near Paris in October 1794. He never intended to conquer the sky. Not any more than the French field marshal who used to praise the courage and obedience of his soldiers by claiming they would 'storm the Moon, should I express the desire thereof'. On the other hand, in 1756, one of Napoleon's predecessors, Frederick II of Prussia, wished to reward one of his subjects, Aul Jürgens, by granting him ownership of the Moon. Why has one of his descendants, named Martin, eventually claimed it in 1999? Because a wind of commercialization of the Moon had started blowing in the twentieth century . . .

Between the 1930s and the first decade of the twenty-first century, Virgiliu Pop identified at least twenty-five analogous claims regarding the Moon, Mars, and other celestial bodies, sometimes even the entire sky. Some claim property deeds or wish to purchase extraterrestrial parcels, others demand rights when a space probe or crew land on the (so-called) extraterrestrial property of a human[13]. These claims are mainly addressed to NASA, the President of the United States, and the United Nations. James Thomas Mangan, for example, did not hesitate to declare that his home was the center of the universe and of the realm of Celestia, for which he handed out the responsibilities and honors to members of his family. On April 6, 1966, the town of Geneva, Ohio solemnly declared owning the Moon. Similarly, Earth

---

13. *Cf* Virgiliu Pop, *Who Owns the Moon? Extraterrestrial Aspects of Land and Mineral Resources Ownership* (New York: Springer, 2009).

people can find property deeds in cereal boxes or at the back of scientific association membership cards. They can also claim to be the owners of a lunar probe in proportion to their tax contributions to space programs.

When potential buyers reply to acquisition offers, they are often immediately warned by the 'sellers'. For instance, Barry McArdle, founder of the Lunar Development Corporation, made it clear that one would have to be a fool to sell the Moon, and also to purchase it. He refrained from guaranteeing his clients supplies of water and air, simply ensuring them . . . a good laugh! Robert R Coles, after having been director of the Hayden planetarium in Manhattan, went into selling plots of lunar land, in the mid-1950s. He never hid the fact that it was nothing but a joke, a prank. Around the same time, the Boston science museum offered the public to buy stars and planets in order to contribute to the construction of a planetarium. The people in charge of the museum made sure to point out to the new owners that they wouldn't be providing them with means to reach the place or to visit their purchase. Their example was followed and copied by other scientific institutions and museums. The Great Martian Land Sale and The Mars Fever Week, organised in 1982, in Boulder, by the planetarium of the University of Colorado to overcome a serious financial crisis, were never presented as anything other than as a 'gag', a mere joke that, nevertheless, had an unexpected nationwide impact. How not to applaud the 'teenage rebellion full of beauty and eccentricity' triggered by Romanian highschool kids in Buhusi, who declared owning Mars and undertook to sell plots of it to finance construction of a new sports hall? What should we make of the acronym of the company created to put plots of Uranus up for sale: R.E.C.T.U.M. (Real Estate Commission & Trust of Uranus Management)? Amid all this, on the market of lunar affairs, the best salesperson is without a doubt Dennis Hope, from whom I bought my own lunar plot. In March 2004, in his capacity as 'Head Cheese' (since Anglo-Saxon culture likes comparing the Moon to a cheese), Hope presented in Las Vegas the constitution of a galactic government. He thereby was trying to give his 'movement' a political twist to engage in a dialog with the American administration. Indeed, in January of that same year, George W. Bush had announced a new space exploration program. Thus, Hope offered NASA to rent from him a piece of land of 30,000 acres for a period of 400 years. If this may have put a smile on the faces of American leaders, those of the emerging space power represented by China seem to have not appreciated Hope's initiative to open an embassy in Beijing. In October 2005, his commercial license was retracted, and he was ordered to reimburse his Chinese clients.

In spite of these misfortunes, Hope didn't lose any of his commercial optimism, neither did other entrepreneurs in the same field. For instance, at the same period, Trans Orbital, a Californian company founded by Dennis Laurie, came up with the idea of building a space probe called Trail Blazer, to send it into orbit around the Moon and have it take pictures of the places where the crews of the American Apollo missions and the Soviet probes had landed. Then, they were to land it on the Moon. According to the advertisement broadcast by Trans Orbital, this was to be the perfect opportunity for humans to leave something on the surface of the Moon, such as a message of 300 to 9,600 characters, a business card, or even a personal object . . . at a cost of $2,500 per gram. Today, in space, there is no trace of this audacious marketing program, as Trail Blazer was never launched. However, on Earth, there is. Those who had reserved a ticket to the Moon received flashy certificates engraved on gold leaf.

For space law specialists, all these enterprises are obviously 'pure fantasy', 'intellectual fraud' (Armand Kerrest) or worse, 'scams' (Gabriel Lafferranderie). Should one, therefore, go so far as to speak of a fool's bargain? In any case, matters become more serious when an actual exploration of space becomes possible and its exploitation conceivable.

## Time for flags

The first decade of the space enterprise was marked by what is commonly called 'the race to the Moon'—a race that was won by the United States in July 1969. At that time, there was no question of purchasing or claiming ownership of the Moon. It was first and foremost a matter of national pride and flags.

The flags planted on the Moon by the astronauts of mission Apollo caused many heated discussions. Indeed, in its second article, the Space Treaty stipulates that 'Outer space, including the Moon and other celestial bodies, is not subject to national appropriation by claim of sovereignty, by means of use or occupation, or by any other means'. Yet this treaty was implemented on October 10, 1967, that is one and a half years before the first landing on the Moon. Consequently, how is one to interpret Neil Armstrong and Buzz Aldrin's gesture, on July 21 1969? Was it an affront to the Space Treaty, a claim of ownership and sovereignty on behalf of the United States? Before the Apollo 11 mission even took off, the issue of the flag was already keeping NASA deciders busy, and it continued to do so after it returned to Earth, only on a broader scale. The amendment approved by the United States on November 18 1969, put an end to the controversy:

> The flag of the United States, and no other flag, shall be implanted or otherwise placed on the surface of the Moon, or on the surface of any planet, by the members of the crew of any spacecraft making a lunar or planetary landing as a part of a mission under the Apollo program or as a part of a mission under any subsequent program, the funds for which are provided entirely by the Government of the United States.

And the text went on clarifying 'This act is intended as a symbolic gesture of national pride in achievement and is not to be construed as a declaration of national appropriation by claim of sovereignty'.[14] In other words, the American astronauts' gesture was to be interpreted like that of any mountaineer who plants his flag on the summit he has just reached. Even against the backdrop of the Cold War, the victory of the United States over the Soviet Union put an end to this competition, this 'race to the Moon', on July 21 1969. It was truly a matter of supremacy, sovereignty, and national pride, and not of ownership or, even less, exploitation.

The American amendment of 1969 defends the symbolic aspect of planting the national flag, but it does not omit to point out that the Apollo missions are financed by the state, meaning by the American taxpayers. This statement could serve as an argument to defend some kind of protection of the landing sites on the Moon. For example, without strictly claiming ownership of them, the winners of the race could nonetheless demand that these sites have their access regulated,

---

14. https://books.google.fr/books?id=PZveCwAAQBAJ&pg=PA18
0&lpg=PA180&dq=november+1969+moon+flag&source=bl&
ots=jJND1m-0oQ&sig=ACfU3U1g3mhaw0v_ITN-gnjXVSuiwPBanQ&
hl=fr&sa=X&ved=2ahUKEwjs-N6bg7HiAhVa8uAKHSXVAD4Q6AE
wEnoECAkQAQ#v=onepage&q=november%201969%20moon%20
flag&f=false.

or even forbidden, as if they were sanctuaries. Let us not forget that access alone is already a key to development; today, we can mention the examples of access to data, resources, results, etc.

Exploiting a place can also come across as a way of seizing it, either temporarily or not. For instance, although no one can claim to own an orbital location around the Earth, everyone can use this common resource as long as he or she abides by international regulations and procedures. States keep under their jurisdiction, their control, and their responsibility all the objects they have sent into space, or that have been sent from their territory or by their nationals. The International Telecommunication Union (ITU) is in charge of managing how frequencies are attributed. Thus, everything is a matter of common resources, common management, and agreement to cooperate. Nonetheless, one must admit that the use of locations and frequencies can lead to a certain form of (at least temporary) appropriation. This possible interpretation of exploitation practices cannot be overlooked, which is probably why NewSpace companies interested in mining resources in space avoid using the term 'exploitation'.

### The new space rush?

Every time a territorial or commercial expansion takes place, or a new power or knowledge is acquired, one must ask how the associated benefits are going to be shared. How could NewSpace evade this obligation when it claims to offer its clients a new world and to open up new opportunities to GAFA entrepreneurs? The issue is all the more relevant as space law has already provided its own answers to this, even in the very first paragraph of the first article of the 1967 treaty mentioned earlier: 'The exploration and use of outer space, including the Moon and other celestial bodies, shall be carried out for the benefit and in the interests of all countries, irrespective of their degree of economic or scientific development, and shall be the province of all humankind.' Every time the notion of province (in French, *apanage*) is evoked—and even more so when it is invoked—it doesn't fail to spark debates, including among jurists. In France, *apanage* used to refer to the portion of royal domain attributed by the king to his younger sons or brothers, and returned to the king's domain if its holder died without leaving behind a male inheritor. While the term comes from the royal regime, in the present context it does not designate property, but rather a field of activities, namely space exploration and use. Thus, the 1967 treaty juxtaposes two principles. On one hand, space must be explored and used with respect for the good and in the interest of all countries. Without being a finality in itself, this condition is accepted from the start by all signatories of this text. On the other hand, it is admitted that space exploration and use belong to humanity as a whole, meaning to all members of the human species. Nothing can deprive them of it, and they may even claim rights and means to it.

How have these principles been interpreted and implemented over the past fifty years? As usual, the scientific community probably remained closest to the spirit and terms of this article of the 1967 treaty. The data gathered during many space missions, the conclusions drawn, and the hypotheses elaborated have been published and shared for the benefit of all scientists and, as far as communication and education enabled it, for all publics. Conversely, the military use of space has remained confined to a set of codes and conducts specific to this domain, including in terms of technological transfer, whether the goal is to ensure space dominance or to establish space power. In such contexts, international treaties seem to have very restricted power. And between these two extremes, commercial activities have been growing continuously, thanks to organizations that were first created by public authorities, then by private companies. Here we must mention the noncommercial initiative of the international Charter 'Space and Major Disasters' described earlier.

What can actually be expected from NewSpace? Business as usual? Of course, I will leave aside the contracts signed by the military staff with companies such as SpaceX. As for the rest, there is no shortage of debates, especially when studies on asteroid mining operations or space tourism promise to offer considerable profit. Voices are rising to demand that the future benefits be shared as widely as possible, in the name of the spirit of space law, and to suggest that a system of concession and adjudication be set up for the benefit of the United Nations, in the name of humanity as a whole. But does the UN dispose of the necessary authority and sovereignty to receive royalties? Henceforth, it seems utopian to envisage sharing spatial resources and NewSpace benefits fairly, meaning it sounds like pure fantasy, evidently impossible. One would need a great deal of optimism to imagine that space could escape the rule of 'first-come, first-served!'. Unless space manages once more to serve as a mirror for humans' dreams, the idyllic ones as well as the nightmares, and to help people be more realistic about their fate on Earth, to take it into their own hands.

I have previously mentioned the works of the Hague International Space Resources Governance Working Group, dedicated to future space resource exploitation. It is imperative to encourage a juridical reflexion that develops alongside a technical, socio-economical evolution, or even precedes it, rather than 'catching-up' with it. That being said, such an endeavor doesn't exclude the possibility of simultaneously delving into an ethical reflection. On the contrary, it demands it. Indeed, to what extent are such reflections limited, burdened by the fact that they remain within the frame of a world-representation that has now been proved to be outdated? Or, to be more specific, I wonder whether such reflections don't reveal how much the underlying world-representation is ill-adapted to present-day space law (see 'Chapter 4: Law is a mirror of space ethics—New limits').

We can measure the effects and the damage caused by our activities in space. Every day, we update the map of space debris orbiting the Earth. We have images of the sites where our probes have impacted other planets. We are able to estimate the amount of radioactive elements and the population of microorganisms that we have 'dumped' in space or on other planets over time. But how could we measure these consequences on the scale of the universe? They seem so tiny! On the other hand, we have to question ourselves about the consequences of space exploration and exploitation projects for the Earth itself and for its inhabitants. Spatial mining resources should first serve the purpose of further exploring space, but what will happen when they are brought back to Earth? What economic, political, and social consequences can we already foresee? History, especially the history of globalization, abounds in teachings which we should consider. We mustn't forget that 'gold rushes' transform the places from where the pioneers left as much as those where they settled. Space ethics cannot leave the Earth aside.

## Should space be explored at any cost?

On July 15, 1969, the Reverend Ralph Abernathy showed up at one of the entrances of Cape Canaveral (called Cape Kennedy in those days) a few hours before the astronauts Armstrong, Aldrin, and Collins took off for the Moon. Together with a group of activists, Martin Luther King's friend started protesting against the cost of the Apollo program, estimated at approximately $25,4 billion at the time (the equivalent of $167 billion in 2017). The flyers they were handing out to the crowds gathered around the American launching site were eloquent. They described the minorities' misery in the USA and the lack of progress in the fight against racial inequality, while the race to the Moon (and perhaps soon to Mars) was being allocated exorbitant budgets. Tom Paine, director of NASA, sent the following message to Abernathy and his friends: 'If we could solve the problems of poverty by not pushing the button to launch men to the Moon tomorrow, then we would not push that button!' Did Paine really mean what he said, or was he simply being skillfully hypocritical? Was Abernathy too quick in ordering the demonstrators to withdraw? Or did he really admit that the racial issue in his country could not be limited to a few budget lines—as astronomical as they may be—but depended first and foremost on a slow and deep evolution of people's mentality? Forty years later, Barack Obama was elected President of the United States of America and, that same year, Charles Bolden became the first Afro-American to be appointed the head of NASA. This shows that many things have changed in the country of the Apollo mission heroes. But at what price?

**For a cup of coffee**

When they demonstrated at the entrance to Cape Canaveral, Abernathy and his comrades were not only denouncing their country's social inequalities and the estimated amount of the budgets allocated to the Moon race. The American space program only served them as an alibi to question the political line and choices adopted by their so-called democratic society. Does space not usually come across as a government fiat, escaping any real debate on the scale of society, of the electors? Besides, this assessment does not apply only to the United States. In most if not all countries having a space policy, it is rarely if ever discussed in public. Yet when they are questioned on the subject, citizens usually express their concern regarding the cost of space activities, deemed excessive. They often wonder whether it wouldn't be more relevant to make use of public funds to reduce social inequality, improve education, build hospitals . . . or solve the issue of world hunger.

Actually, when questioned about the cost of their programs, players in the space field most frequently answer by putting forth two arguments: that of the economic spin-offs and that of comparison. The first seems obvious: one must take stock of the jobs created, the incomes ensured, and the exports provided by the national or international actions of the space industry. In Europe, it is commonly pointed out that one euro invested yields twenty euros in terms of economic benefits. The second argument is that of 'the cinema ticket or the cup of coffee'. The European astronaut Thomas Pesquet resorts to it when he is questioned about the cost of the Proxima mission. He notes that Europe's annual participation in missions onboard the International Space Station costs each European citizen the equivalent of the price of one cup of coffee. Similarly, the whole of public investments in space-related activities costs each French person the equivalent of three cinema tickets per year, i.e. approximately thirty euros.

What should we make of such an answer? First, it shows that, with moderate investments at the level of each inhabitant (a cup of coffee, a few cinema tickets), we are able to achieve literally exorbitant, extraterrestrial projects, such as conceiving and launching rockets, supporting the needs of astronauts in outer space, sending probes to Mars or onto comets, etc. This is not to be looked down on, and I would even add that the price is very reasonable to show what today's nations are capable of accomplishing when they agree to share their knowledge, skills, and human and financial means. But let's not be naive. Even if we don't measure their extent, we know that space programs involve industrial stakes, complicated interplay between different powers, interests, and influences. Space is not exempt from any of humanity's defects, even the most shameful. I understand the Afro-American minorities' revolt at the time of the Apollo missions, as well as the indignation which could legitimately be felt by other minorities or territories that are too often neglected or ignored by today's societies. I fully grasp the message of those who address us space people, accusing us of 'costing too much to society' and expressing how it would be fairer to dedicate our budget to education and social matters. However, one must be careful not to adopt an excessively reductionistic or fragmentary point of view. The price of a cinema ticket or a cup of coffee could indeed support a program to build a hospital or a school, but it could also serve to finance the renovation of a work of art, the construction of a stretch of highway, or the purchase of an aircraft-carrier or fighter plane. So, what should we choose? In July 1969, Paine and Abernathy probably agreed that their respective commitments and responsibilities (pushing on a button in one case, demonstrating in the other) could merely serve as opportunities to question their entire society about its priorities and common objectives. But are we even capable of addressing such an issue?

To come back to the subject of space activities, specifically of space exploration, we must point out an interesting distinction between price and dignity, introduced by the philosopher Immanuel Kant. In his *Groundwork of the Metaphysic of Morals* (1785), he wrote 'In the empire of ends everything has either a **price** or a **dignity**. What has a price is something in the place of which something else, as an *equivalent*, can also be placed. What, on the other hand, is elevated above all price, that has a dignity'. To answer the frequent questions about the cost of space activities, it is indeed common to report what space budgets were invested by states by giving numbers and ratios as well as by comparing budget lines and public programs. The image of the cup of coffee or the cinema ticket involves the same method. Regardless of how one puts it, it is obvious that, from the standpoint of cost, space activities pertain to the field of big or even huge investment and funding programs, committed to and undertaken by states. No more, no less. Henceforth, why deem them too costly? Should this reaction be considered as an effect of the obvious process of trivialisation affecting the most useful space tools? Humans have been so quick and so good at getting used to having satellites fly over their heads (for purposes of communication, observation, and positioning) and using their services, that they have ended up forgetting about them. They even seem surprised that satellites still need to be conceived and built, launched, and kept in good working order to guarantee the multiple remote services provided by these machines and enjoyed day after day by the Earth's inhabitants. These go from weather predictions to navigation assistance, from international communication networks to observation and surveillance. Discrete, invisible to the naked eye, these satellite systems effectively have a cost, but also a price, in the Kantian sense of the term. They have a price which could or should be compared and estimated within the broader context of accomplishing a mission, obtaining a result, achieving an end, but also with regard to equivalent terrestrial services when they exist. Is it a matter of transmitting information as widely as possible? The benefits and costs of satellites must be compared to those of optic-fibre cable, according to the accessibility for the populations involved. This comparison actually leads to associating the two methods rather than opposing them, or setting them in competition with one another. It would be possible to proceed in the same fashion for all space techniques and operations offering us their services. Thus perceived and estimated, we can indeed say that space has a price, but that it isn't exorbitant.

Space has led to the development of another sector of activities, that of exploration, meaning space telescopes, planetary probes and, to a lesser extent, manned flights. Whether the cost of these missions is considered modest (a space telescope able to detect extrasolar planets costs several million euros) or whether it is more costly (an exploratory mission on Mars, carried out by a robot or an automated probe, is presently estimated to cost several billion dollars; the Rosetta mission studying the Churyomov Gerasimenko comet cost one billion euros), one has to admit that such space activities have no equivalent and must, therefore, be measured or evaluated other than by their cost. Exploration deserves to be assessed by means of the second notion suggested by Kant, that is dignity. This position is neither exaggerated nor inappropriate. What would have become of our species, had it not, since the dawn of its time, undertaken highly dignified enterprises to satisfy its natural curiosity and perhaps innate propensity for exploration? Increasing our understanding of the universe, life, and their origins, braving lands thus far ill-known or completely unknown, taking the risk of shaking up ideas, theories, certainties ... What would become of us if we decided to stop letting ourselves be carried, influenced, and inspired by such tendencies? Answering these questions goes way beyond mere armchair philosophy or the argument of the cup of coffee. It requires delving into the very foundations of the human condition: the history of humanity, its cultures and societies, its values, projects, and hopes. As it has in the past, space could, in the future, still engender deep philosophical questions.

## In the face of death

The space enterprise has taken many human lives. Before anyone had even envisaged sending men into space, many prisoners of Nazi concentration and detention camps lost their lives while building V-2 rockets, developed by Wernher von Braun and the Peenemünde engineers by the Baltic sea. Once the United States and the Soviet Union had initiated their space race, many more died 'on the ground', while perfecting rockets or launching them. Hereafter, a series of examples: Marshall Mitrofan Nedelin, in charge of the Soviet strategic forces, died alongside more than one hundred people when an intercontinental R-16 missile exploded, on October 24, 1960, in Baïkonour. Fifty-six inhabitants of a village in the vicinity of Xiang (the launching base of the Chinese Longue Marche rockets) when one of the rockets failed on February 14, 1996. Twenty-one Brazilian engineers and technicians during an explosion on the launching pad of Alcantara on August 22, 2003. Three American technicians testing the engine of SpaceShipTwo, in Mojave on July 26 2007. One of the pilots of the Virgin Space Ship Enterprise, again in Mojave on October 31 2014. The community of astronauts, cosmonauts, and taikonauts has also paid a heavy tribute to space. Since the beginning of manned flights in 1961, over one hundred and fifty men and women have gone into space, and twenty-two of them have perished accidentally, either during a mission or during their training. Space shuttles, conceived and used by Americans, have turned out to be particularly dangerous. According to Jeffrey Bell, it seems to have been six times more perilous to take a seat in one of them than in the cockpit of a B-24 Liberator bomber flying towards Nazi Germany in 1944. If the value of such an estimate can always be put into perspective, or even criticised, it nevertheless coincides with Richard Feynman's statement after the Challenger disaster. The Nobel Prize physicist had compared deciding to launch Challenger to playing Russian roulette: the fact that the first shot didn't 'go off' doesn't give any indication of subsequent events.[15]

---

15. 'Personal observations on the reliability of the Shuttle', in an addendum to the Commission of enquiry into the Challenger accident (http://science.ksc.nasa. gov/shuttle/missions/51-l/docs/rogers-commission/Appendix-F.txt).

## Two different risk cultures

At the beginning of the history of manned flights, Americans and Soviets similarly followed three basic principles:

- the safeguard of the people on the ground was to come before that of the crew,

- the crew's solidarity was to ensure that one would always try to save the entire crew rather than any particular one of its members,

- and the concern for safety was to be primordial throughout each step of a mission, with specific procedures and means adapted to each phase of flight, including on the launching pad and during the flight ascent stage.

One illustration of this third principle is the installation of a mini-rocket above the inhabited capsule so that, in the event of an accident during the launch, it can be detached from the rest of the rocket and, thus, save the crew. The Russian Soyuz is still equipped with such a system, just like the American spacecrafts Mercury, Gemini, and Apollo used to be. In fact, Soyuz attributes several successful rescues to it. Another example is the choice of a return to Earth via a ballistic free fall, ending on land or in the sea. The maneuver is more brutal than a gliding return, but it is less perilous. So, was it the almost unexpected success of the Apollo program—even including the Apollo 13 accident ('a successful failure', as it was coined back then)—that led NASA to reconsider its approach to safety and to reassess its principles for the conception of its space shuttles? The American Space Agency has now completely ruled out the possibility of saving the crew during the solid-booster propulsion stage. During those hundred and twenty seconds, the astronauts are at the mercy of technology, with the now infamous consequences experienced by the Challenger crew members. Conversely, when they built their own shuttle, the Soviets remained faithful to their tradition. Although Bourane (that was its name) oddly resembled its American counterpart, it still included ejection seats and, unlike the system chosen by NASA, it didn't ensure any propulsion itself during take-off, meaning that, in the case of an accident, it could easily have been separated from its huge carrier rocket, Energia. However, Bourane only did one test flight before it was put away in a hangar; what's more, it was an automated flight, that is without any cosmonauts onboard!

At the time of the space race between Americans and Soviets, fully financed by public funds, it was commonly admitted that astronauts sent to the Moon had only one chance in two of safely returning to Earth. But those days are over. Nonetheless, several countries are currently studying programs to return to the Moon or to plan manned flights beyond, meaning that the possibility of accidents or failure cannot be ruled out. Engineers are still as unrelenting as in the past, and astronauts are still as brave, so one cannot say that the stuff of which heroes are made has become obsolete, nor that it should be locked away in space museum showcases. Failure is not an option that goodwill alone, or intelligence, or even extreme precaution can rule out completely and permanently. Those in charge of preparing space programs must continue to anticipate the possibility of an incident, an accident, or even a human disaster. Space exploration is still likely to exceed the limits of our capabilities, not only those of our knowledge and technology, of our political and financial assurances, but also those of the value of human life. Is it reasonable to develop and defend programs of manned flights to Mars although we know how dangerous prolonged exposure to cosmic rays can be for crew members, and we still haven't found any efficient ways to protect them against such risks? Is it reasonable to send humans on a mission so long and distant from Earth that any transmission between the crew arrived on Mars (or in its vicinity) and mission control would require forty minutes? Those who stubbornly defend the need to launch a mission to the Red Planet without further delay seem to minimise such constraints and dangers, while, on the other hand, their contradictors emphasise them and sometimes exaggerate them. Which of the two parties is right or wrong? Regardless of what techniques are eventually used, if a mission to Mars ever takes place, it will inevitably be a dangerous operation. Courage will not merely be seen in how the future deciders will set up such a mission, or how astronauts will partake in it. Courage will also consist of foreseeing the most fatal outcome and preparing for it. Here, I am thinking of the task entrusted to William Safire on July 18, 1969, three days before Armstrong and Aldrin landed on the Moon. He was asked to write the speech that President Nixon would have to give in the event the two astronauts would remain trapped on the surface of the Moon. The counsellor of the President of the United States wrote the following text:

IN EVENT OF MOON DISASTER: Fate has ordained that the men who went to the Moon to explore in peace will stay on the Moon to rest in peace. These brave men, Neil Armstrong and Edwin Aldrin, know that there is no hope for their recovery. But they also know that there is hope for humankind in their sacrifice. These two men are laying down their lives in humankind's most noble goal: the search for truth and understanding. They will be mourned by their families and friends; they will be mourned by their nation; they will be mourned by the people of the world; they will be mourned by a Mother Earth that dared send two of her sons into the unknown. In their exploration, they stirred the people of the world to feel as one; in their sacrifice, they bind more tightly the brotherhood of man. In ancient days, men looked at stars and saw their heroes in the constellations. In modern times, we do much the same, but our heroes are epic men of flesh and blood. Others will follow, and surely find their way home. Man's search will not be denied. But these men were the first, and they will remain the foremost in our hearts. For every human being who looks up at the Moon in the nights to come will know that there is some corner of another world that is forever mankind.[16]

Fortunately, this speech was never delivered, and President Nixon was able to have a conversation with the three astronauts on the following July 24, once they had returned safely from their lunar journey. NASA had achieved the challenge issued by John Kennedy barely eight years earlier, '(. . .) before this decade is out, of landing a man on the Moon and returning him safely to the Earth'.

### Envoys of humankind

In the meantime, the Space Treaty has conferred to astronauts and cosmonauts the title and mission of 'envoys of humankind'. This expression gives the whole space enterprise a singular humanistic dimension that must be associated with another juridical notion, namely that of 'province of humankind' applied to space exploration and use. Envoys of humankind, province of humankind: both expressions are particularly inspiring and imprecise. Skeptics point out that

---

16. *Cf* http://gawker.com/5369364/william-safires-finest-speech.

granting such a title and mission to astronauts, cosmonauts, and tai-konauts, and giving such a status to the space enterprise should not happen unless humankind has at least acquired an awareness of itself. This, however, is something that it cannot really pretend to have, even today, fifty years after the Space Treaty. Nonetheless, they are met with the following answer: is it not during such high-risk enterprises, led at the edge of its knowledge and capabilities, that humankind can become truly aware of itself and of the globalisation process which it has entered, whether its members are aware of it or not, whether they are actively participating in it or not? When, in the 1960s, individuals were declared envoys of all humankind and their enterprise its prov-ince, such a statement necessarily rested on an ideal and a horizon that could not be approached without facing dangers, without involv-ing, assessing, and mastering many risks. Those who wrote up the fundamental texts of space law knew this, just as they knew they were acting against political and ideological threats that were abundant in those days, as they are today, including threats of nations or private companies appropriating space or weaponising it.

No one had ever imagined that space could be conquered without anyone having to pay a price for it. In the West, the ancient myth of Icarus is the most eminent proof, if not the oldest, of such awareness. Those who decided to confer upon astronauts the singular title of 'envoys of humankind' were aware that these men and women would sometimes be sent to their death, without any warlike motive being the cause of their demise. Death is one of the conditions of space exploration, not in the sense that it is a necessity, but in the sense that it is a possibility inextricably linked to its very nature. Refusing it from the outset comes down to refusing the enterprise itself. Yet the question remains: how can a group of institutional leaders, a scientific community, a civil society, a state accept to put not only consider-able scientific, technical, and financial means at risk, but also people's lives, knowing that their freedom, their independence, their safety, or even their prosperity are neither being questioned nor jeopardised?

The possibility of a disaster, the option of death reminds human-ity that not everything is allowed, that not everything is conceivable, and that it has to make choices, meet decisions, and commit without knowing in advance what the outcome of its enterprise will be. The days are over when the gods used to decide on humans' fate. Hence-forth, when they take their own existences and fates into their hands, humans must beware of themselves.

# Chapter 8
## Space of Values, Values of Space

'Has the discovery of America been useful or detrimental to the human race?' Preoccupied by the question of slavery, Father Raynald presented a thesis dealing with this question in 1787, in front of the Lyon Academy. Sixty years after the beginning of space exploration, and at a time when the 'space rush' is a highly topical issue, we are entitled to ask ourselves the same question as the French ecclesiastic did back then: have the exploration, conquest, and exploitation of space been useful or detrimental for humankind so far, and what will they lead to in the future? Seeking to answer this is part of the task of ethics, but pretending to have found satisfying answers would be pretentious and even erroneous. The contents of this essay should suffice to prove this point. Nonetheless, it would be just as wrong to deny the progress that the space enterprise has enabled humankind to accomplish and the teachings it has conveyed. Alongside what we have learned in the fields of science and technology, I would like to point what we have also learned in terms of values—human values, to be more specific.

Dealing with space provides us with a unique way of expanding humanity to the dimensions of the universe and concentrating the universe to the dimensions of humanity. Thereby, space truly participates in the process of humanisation of our species, as well as in the general, global, integral development of our species, at the individual and societal levels. This development requires us to resort to driving forces, props, points of reference, in other words, values inherited from the past. Yet at the same time, this process is favorable to the emergence of new values. I will not delve into the moral values that are most commonly accepted by past and present philosophers or human cultures (respect, mutual assistance, benevolence, brother-hood, qualities of welcoming and listening, etc). Instead I will focus on those that seem to me more specifically connected with the space enterprise, for instance, the principle of common heritage, the notion of 'space of possibilities', and the reality of the boundaries of time. Let me clarify my point.

## Common cause

On several occasions throughout this essay, I showed how space activities have contributed to the dynamics of globalization, which is one of the distinguishing features of modern times. When roads need to be built around our planet—both real roads (on land, on the sea, or in the air) and virtual ones—satellites now help engineers, construc-tors, and navigators of all types in a precious though singular way. The images provided by instruments orbiting the Earth or located millions of kilometers away can be shocking, especially when astro-nauts or astronomers (namely Carl Sagan) add to them the weight of their words. Under the overview effect, their sentiments describing the fragile beauty of the Earth, its atmosphere, its forest cover, or its vast bodies of water have underlain the birth of ecologist movements and have, up to this day, served to raise awareness in favor of the earthly environment.

Without anyone having really seen it coming, the Earth has become one of the prime destinations and goals of space missions, but still, humanity remains their first and foremost raison-d'être. After the success of Apollo 11, many people exclaimed 'WE went to the Moon!' At that same period, space law conferred astronauts the sta-tus of 'envoys of humankind'. Moreover, when he discovered the 'Pale Blue Dot', that image of Earth taken by Voyager 1 in 1990, Carl Sagan exclaimed 'That's us!'. Experiencing the singularity of our planet, and even its solitude at the heart of our universe, leads us above all to take the measure of the singularity and solitude of our species. And this experience also raises a sense of unity: 'We are all one!' In parallel, the exponential development of means of communication and social net-works is enabling the appearance of federations of various types. In France, after the terrorist attack at the headquarters of the magazine Charlie Hebdo, a movement of protest and solidarity grew around the motto *Je suis Charlie!* (I am Charlie!). This kind of identification has taken on such proportions that each of us can even say 'I am the crowd' (Jacques Blamont). Here we face the double process of expansion-concentration which I was refering to earlier, in the context of space exploration, and which I consider to be a major contribution to our humanization.

This being said, space does not only offer us an experience of our unity: it also invites us to strive towards a common cause. This is what I wanted to highlight in my analysis of the notion of common heritage. This cornerstone of space law is not merely a jurist's whim. Or rather, it would be, if it were only interpreted in terms of non-appropriation. In such a case, it would be ridiculous to defend celestial bodies, including their exploration and exploitation, as part of humanity's common heritage. How could humanity possibly pretend to claim ownership over space while it is totally unaware of its limits? On the other hand, if claiming and defending space as humanity's common heritage entails and implies a shared responsibility towards the celestial bodies that are progressively being explored or even exploited by humans, then this notion deserves being taken seriously, and its application to human enterprises should be thoroughly envisaged and fought for. It is true that the projects of NewSpace entrepreneurs shake up the interpretations and applications that had thus far been commonly accepted. Yet, there is nothing disastrous about this: it simply forces us to step up to the challenge of adapting our actions, commitments, mindsets, and behaviors, in short, of making our ethics evolve, but without necessarily forgetting or forsaking the values introduced at the time when space activities effectively began. Whether it is a matter of pursuing space exploration (and, today, of preparing missions towards the Moon or Mars) or of exploiting the Earth's orbits even more intensively, we must imperatively think in terms of common heritage, shared responsibility or, as I have mentioned earlier, sustainable development—in the name of a single, shared concern for our humanity.

This being said, we must not let this experience of community and globalism fascinate us to the point of deceiving us. Both our species and our planet still have boundaries, and this will always be the case. Yes, seen from space, natural and artificial borders between humans remain visible, but this is not a regrettable fact which we should learn to deal with, ignore or necessarily oppose. On the contrary, it is a necessity, an imperative for the survival of our species. If, as members of a same species, we feel the urge to measure and feel our oneness, to defend that which unites us and brings us together, we are just as much in need of discovering and acknowledging that which differentiates us. We cannot do without borders, that serve at the same time as limits to our freedom and to our personal ambitions, and as anchors, bases upon which we can establish bonds, realistic and reasonable connections between each other and between our societies. In space, this reality is illustrated by the double dynamic process of competition and cooperation.

The history of space can be seen and interpreted as a succession of competitive and cooperative episodes between countries, now including private companies too, that have progressively got involved in the acquisition and development of astronautic technologies. However, competition and cooperation are more than keys to understanding history: they serve as real driving forces and values for space. If competition underlay the race to the Moon and can explain the American success of July 1969, cooperation was the fundamental basis upon which the International Space Station (ISS) was built. To this day, the ISS is the largest artificial object ever placed in orbit and most likely one of the most complex objects ever made by a human. Nonetheless, it would be wrong to oppose competition and cooperation too drastically. Though they are driven by different spirits, they do resemble each other on one point: they require a clear definition of a single, common goal, to be reached either first or together. In other words, those who fight in a competition as well as those who undertake a project in cooperation, all work towards a 'common cause', even if they take different paths to get there. As long as we always take care to define or clarify the mainsprings and finalities of our space activities and enterprises, as well as the means to achieve them and the consequences they may entail, our focus will always be to act in full respect of the general interest, keeping in mind the perspective of a common cause, choosing to resort to the dynamics of competition or cooperation depending on the circumstances.

## Space of possibilities

I have evoked the astronomic revolution that took place in the West at the beginning of the seventeenth century. I have also explained how the human person's conception of the sky, and more generally of his reality, was shaken up by the discoveries of Galileo and Kepler, as well as their colleagues and successors. The world is no longer a cosmos, an orderly, harmonious system at the centre of which humankind is confined to the surface of an imperfect and perishable Earth. Henceforth, the world is a universe whose centre is everywhere and whose circumference is nowhere, according to a particularly appropriate philosophical phrase. The landmarks, rules, and alliances of the past have disappeared. Humanity can rely on nothing but itself to build its own destiny. Admittedly, its freedom has not become absolute: its future is still greatly defined by its environment (both predictable and unpredictable), its natural heritage, its cultural traditions, its technical capabilities, its lack of knowledge, etc. But it now finds itself on the threshold of a space of possibilities which it hadn't envisaged yet. Its future is not fated, it is not some path written in advance: it will be, at least in part, what the human person decides to make of it.

Space has a great deal to do with this new paradigm. There is nothing excessive about Hannah Arendt's enthusiastic words, at the beginning of the preface to her essay *The Human Condition*. The German philosopher mentions the first Sputnik, which, in October 1957, 'for some weeks circled the Earth according to the same laws of gravitation that swing and keep in motion the celestial bodies—the Sun, the Moon, and the stars'. She writes of its 'sublime company', although it is 'bound to earthly time'. 'This event, second in importance to no other, not even to the splitting of the atom [is] the first step toward escape from men's imprisonment to the earth'.[1] Yet, becoming aware of this escape was merely the catalyst of a new exploration.

In addition to considering the momentum that has made our species particularly gifted for exploration from the outset (I have dealt with this issue earlier in this essay), one must take stock of the immoderate dimensions of the territory which humankind has undertaken to explore since the first Sputnik. Similarly to the 'infinitely small', or perhaps even more, the 'infinitely big' seems limitless. Rather, to be more specific, the only limits seem to be those imposed by human intelligence, by our knowledge and techniques. Indeed, the field of possibilities is anthropocentric and its dimensions are 'anthropodependent'. That is why, alongside the responsibility associated with

1. *Cf* Hannah Arendt, *Condition de l'homme moderne* (Paris: Calmann-Lévy, 1983), 33–34.

this characteristic of space (clarified earlier, in the paragraph about the notion of 'common cause'), awareness and risk assessment are essential values.

I have shown how the very idea of risk appeared when the cosmic conception of reality ended. Humans became stronger than gods, stronger than destiny, which had thus far been imposed upon them. They accepted the fact that their fate would depend on their actions, on their own initiatives and choices. It is important to keep in mind the statement attributed to Gene Kranz by the director of the film *Apollo XIII,* 'Failure is not an option'. This imperative remains perfectly valid for the conduct of space matters, especially when human lives are directly involved. Nonetheless, Elon Musk's imperative is just as worthy: 'Failure is an option here. If things are not failing, you are not innovating enough'. The whole paradox of innovation, the ambiguity of exploration, rests on the necessity to honor and accept both of these imperatives. Only by associating the two can one hope to open the door to the space of possibilities on the long run and in a sustainable way. I have to think of the answer I often give youngsters, high-school kids or students, when they ask me for advice regarding their orientation towards careers in aviation and space: 'Evaluate your abilities, your possibilities . . . and then, aim a bit higher. Take a risk, but a conscious, reasonable one'. In the fields of space and space ethics, I am convinced that there can be no other solution, for the mere reason that we are only at the dawn of its discovery, on the threshold of immense *terrae incognitae,* boundless unknown territories.

## The verdict of time

While imagining 'scientific marvels' such as the chronoscaphe, the matter transmuter, or hyperspace, thereby becoming one of the masters of science-fiction, Arthur C Clarke effectively participated in the birth of modern astronautics. As a member of the British Interplanetary Society, in 1939, he published an article called 'We can rocket to the Moon—Now!' In 1945, he sent a letter to the editor-in-chief of the magazine *Wireless World,* bearing the title 'Peaceful Uses for V-2', in which he foresaw the launching of geostationary satellites. That same year, the magazine published his article 'Geostationary Satellite Communications': this was one of Clarke's most famous 'predictions'. Paradoxically, ten years after the Apollo 11 mission, this visionary and space aficionado wrote that space travel corresponds to a 'techno-logical mutation that should not have occurred until the twenty-first century'.[2] What a surprise to hear a prophet lament that the future has arrived too quickly, while in his speech of September 12 1962, John F Kennedy had asserted that 'We are unwilling to postpone[3]'! In truth, when it comes to dealing with space matters, it has never been easy to manage time and its constraints, especially when looking forward.

It seems obvious that public and official organizations in charge of space programs have always used a form of language turned towards the future. Each and every program necessarily extends over a time span that includes a deadline and a horizon. From the first draft to the beginning of the technical implementation, or from the launching to the end of the mission, the successive steps often span decades. The institutions in question have never hidden the difficulties that would have to be faced, nor the technological challenges that would doubt-lessly be met. They do not hesitate to evoke the 'high risks' linked to the arrival of a space probe on the surface of Mars, even though no human life is actually threatened in the process. Space is always written in the future tense; however, its usual institutional play-ers run into serious difficulties today when it comes to presenting their programs to political powers or public opinion. These problems arise from the complex, perhaps impossible equilibrium that needs to be found between the apparent repetition of space missions, that is their trivialization, and the exceptional nature of manned flights reserved to an elite (though for how much longer?). I do not believe that NewSpace companies will be able to escape such difficulties and challenges for very long . . .

The fact that humanity is characterised by its cultural singular-ity, explicitly illustrated by the diversity of its artifacts, doesn't mean that it can totally escape nature's reality. The evolution of living beings typically follows two different rhythms: the rhythm of slow, sustained transformations that are barely perceptible, and the rhythm of events such as disasters that suddenly disrupt innumerable forms of life and ecosystems. Our societies and activities are subject to the same constraints of time. Knowing this and accepting it is part of the wis-

---

2.  *Cf* James A Dator, *Social Foundations of Human Space Exploration* (New York: Springer-ISU, 2012), 27.
3.  *Cf* https://er.jsc.nasa.gov/seh/ricetalk.htm.

dom, or rather the ethics, which we need to acquire individually, collectively, and professionally. One of the keys is probably the proper management of the *kairos*, in other words of the opportune moment: as far as space goes, that would mean the opportune moment to confront the unknown. Is it still too early? Or is it already too late? At that precise moment, each human being finds himself alone, facing a choice that must be made. Therefore, the notion of *kairos* is intricately connected with the need to be deeply convinced.

Deep conviction cannot be broken down to a general, quick or superficial impression. On the contrary, it requires scrutinizing each item of the file, submitting to the rigor of reflexion each piece of evidence, each component of the process to conclude or of the procedure to set into motion. Invoking deep conviction, demanding it, is more than a mere formality, a simple work method, or an obvious mental attitude. Deep conviction combines a search for the individual and collective good, and a form of modesty when faced with the reality of facts, the imperfection of knowledge, the responsibility of a trial, the gravity of a decision's consequences. Because it requires associating thought and action, practicing deep conviction is bound to a true ethical position defined by an essential characteristic, even if it seems paradoxical at first sight: collegiality. Now one has to admit that this is a real constraint, because it requires time, patience, and concertation before the decision can be made, that will then be defined as a verdict, 'a human truth of the moment'.[4] To summarise, deep conviction is constituted and inspired by the following requirements, constraints, and spirit: method, reflexion, humility, collegiality, and control over the moment of the outcome. The latter can be a count-down since it implies a time delimited by the experience of the past and the expectations of the future, an 'interval of time entirely determined by things that are no longer and things that are not yet' (Hannah Arendt). This time is that of risk *par excellence*, of something that comes close to the moment of truth, to the verdict of time.

I admit there is a risk in suggesting that these values were born from the experience that we, humans, have acquired after approximately half a century of space enterprise. The risk would be for us

---

4.  *Cf* Fayol-Noireterre, Jean-Marie. 'L'intime conviction, fondement de l'acte de juger (Deep conviction, a foundation of the act of judging)', *Informations sociales*, 127, 7/2005: 46–47.

to limit ourselves to these values and, consequently, adopt a form of dogmatism or fundamentalism. But no, these values are nothing more than assumed and accepted constraints, or rather suggested points of reference. They are meant to inspire us, in other words to encourage us to act and commit ourselves, rather than to feed our timorousness or, worse, our hypocrisy. They must be at the service of an impetus to go forward. On April 12, 1961, as soon as his rocket was airborne in Baïkonour, Youri Gagarine simply announced: 'Poyekhali! We're off!'

# Conclusion:
## Space or the Odyssey of the Future

*Odyssey*: the title of Homer's epic poem, recounting Odysseus' perilous journey, is now commonly used to designate the account of any voyage filled with adventures and, more broadly, any succession of exceptional events occurring in the life of a person or group of people. Thus, the term 'odyssey' is fitting for the space enterprise. Just like the poet's *Odyssey* in ancient times, space abounds with gods, heroes, and creatures so far unheard of; exotic shores, inhospitable places, and *terres-patries* (an expression by Michel Serres, describing the notion of homeland-regions); ruthless fights, impossible challenges, and unexpected alliances. Consequently, we sometimes hear people wonder why space should be burdened with the weight of ethical reflexion. Should it not, on the contrary, be free of all constraints to enable humans to enjoy it fully, to see it as an opportunity to accomplish their personal and collective destinies unrestrictedly, even if that means eventually returning 'on the good Earth' (Frank Borman, Apollo 8), enriched by the wisdom of this new experience?

Indeed, ethics can be perceived as a fascinating odyssey, just like the crossing-over from ignorance or faith ('I can') to reality and effective choices ('I do'), but they are also a burden. They don't only bear the inevitable weight of laws, norms, and codes of good conduct, but also and above all the weight of reflexion. Thinking is a demanding, restrictive, and difficult process, as it leads us to question continuously the 'whys' and 'hows' of our behaviors and choices. Answering these questions involves combining various fields of knowledge and research: science and expertise, history and sociology, philosophy and epistemology, etc. One must possess the suitable and necessary means to conduct a critical analysis, to elaborate an opinion, and to reach a decision. Building such collective intelligence requires time and effort because beliefs and assertions can differ greatly from one individual to another. Nonetheless, triggering reactions, collecting them, and confronting them are ground-laying operations in any worthy ethical process.

As I have pointed out throughout this essay, space must carry on following the ethical process to which it has been committed for some twenty years. In the previous pages, I have described the first stages of space ethics. They include introducing regulations and procedures to manage the use of circumterrestrial space, elaborating agreements to implement the concept of humanity's common heritage, the conditions of exploratory missions to other planets, the spirit imbuing manned flights and space tourism, the governance of the transition from exploration to use to commercialization, etc. New objects of study will not fail to appear in the future, whether they can be foreseen today or not. Analyses will be done and decisions will be made; some will be good, some not. Actually, it is impossible for us to predict anything precise: space will be what our descendants do in it and make of it.

In any case, we must remain confident. As I have mentioned, since *Homo sapiens sapiens* was given that name, he has always practiced ethical questioning, often without even being aware of it. The current challenge is to carry on giving him the means to carry out this task, yet with ever more efficiency, discernment, clairvoyance, and critical distance, in short, with what all cultures qualify as wisdom. Our field of possibilities in space must be better known and better occupied before it can be extended any further.

The space odyssey deserves our interest, not only because it has and will rely on important principles (cooperation, concern for the common good, spirit of innovation), but also because it provides grounds for pursuing the uninterrupted quest which has shaped humanity throughout its history and still contributes to making it what it is to this very day. It is in humankinds nature to concentrate the universe to the dimensions of humanity and to expand humanity to the dimensions of the universe.

# Annexes
## Annex 1: Animals of the shadows

September 19 1783: an aerostat took off from the gardens of the Versailles palace, near Paris. It was a lighter-than-air craft, conceived by the brothers Montgolfier. A sheep, a rooster, and a duck were embarked in a wicker basket. Under the eyes of King Louis XVI, the first aeronauts in history rose to an altitude of 480 meters and, after a ten-minute flight, landed in the forest of Vaucresson, a few kilometers away. The landing was abrupt: the rooster suffered a broken leg. Two months later, on November 21, Pilâtre de Rozier and the Marquis of Arlande performed a twenty-minute flight, West of Paris. Unlike the animals who had preceded them, they were able to feed the fire of their aerostat and remain airborne longer.

November 3, 1957: barely one month after the flight of Sputnik I, the first artificial satellite sent into space, a Soviet rocket took off from the Baïkonour base in the middle of the Kazakh steppe. Onboard was Laïka, a female mongrel found straying in the streets of Moscow and trained by Soviet scientists in view of becoming the first living creature in space. After having orbited the Earth nine times, the temperature inside the capsule increased until it exceeded 40°C. Lacking sufficient protection against solar radiation, Laïka died of heat exhaustion and dehydration.

CNES researcher Michel Viso explains that 'The USSR chose a dog because it was the animal which the Soviets knew best from a physiological standpoint', just like sheep, roosters, and ducks were most familiar to scientists at the end of the eighteenth century. In both cases, the goal was to prove that a living being is capable of surviving in unusual conditions, such as in flight (the duck was then the model) or in weightlessness.

Laïka was not the first guinea-pig of the space conquest. As early as 1947, fruit flies (Drosophila) were placed onboard a V-2 rocket to determine the effects of high-altitude radiation on living organisms. Around that same time, rats, mice, and rabbits were used in similar experiments onboard balloons. Between 1948 and 1951, the United States sent six rhesus macaques (from Albert I to Albert VI) onboard rockets destined to reach the frontiers of space. None of them orbited the Earth and only the sixth survived, although only briefly after having reached an altitude of 72 kms. In July 1951, before Laïka's feat, two Soviet dogs, Tsygan and Dezik, had survived a suborbital flight at an altitude of 100 kms. Youri Gagarine's flight on April 12 1961, was preceded by the flight of bitches Belka and Strelka, who returned from their journey alive . . . and thereby became legends.

In the United States, monkeys carried on serving in the shadows, paving the way for the first manned flights. In May 1959, Miss Baker, a squirrel monkey, and Miss Able, a rhesus macaque, made a suborbital flight onboard an intercontinental missile. In January 1961, Ham was the first chimpanzee to go into space thanks to a rocket of the Mercury program. In November of that same year, Enos, another chimpanzee, achieved an Earth orbit, but died six months after returning from its mission . . .

After Gagarine's inaugural flight and given the increasing mastery of manned flights, why do space agencies still resort to animals? Simply because a human being 'is not an ideal guinea-pig, for one cannot perform the same in-depth examinations on a human organism', explains Dr Viso. That is why scientists continue sending appropriate guinea-pigs into space, choosing them according to the aims of their research. After Hector the rat and Félicette the female cat, sent into space by French rockets in 1961 and 1964, two mummichogs (killifish) and two spiders, Arabelle and Anita, tested their respective abilities to swim and weave a web in zero gravity onboard the American Spacelab in 1973. Shortly before that, Soviets had placed turtles in the same weightless environment, making them undertake the longest animal mission in space thus far: 90 days. The issue of reproduction in outer space does not only interest the media thirsting for sensational news, but also researchers. They resorted to salamanders onboard the ISS in 1996 and 1998, and to geckos in 2004 (whose capsule, Photon-M4, unfortunately went missing). They successfully inseminated mice with semen obtained from animals that had sojourned nine months on the ISS. Muscular atrophy and bone loss, which are both possible consequences of long stays in outer space, have been studied on animals as diverse as mice, Mongolian gerbils, lizards, and snails.

One can say that, up to now, animals have served as humans' emissaries in space exploration rather than as mere guinea-pigs. Will we resort to them again in the future to go further into space and for longer periods of time? Ethics can help us develop an answer to that question.

# Annex 2: Star Wars?

Many experts wonder whether it is appropriate or not to speak of the militarization of space. Since the 1967 treaty announced that the Moon and other celestial bodies could be used only 'for exclusively peaceful purposes', and since that same treaty strictly forbade 'testing weapons of all types and executing military maneuvers' (but in parallel merely forbade 'placing in orbit around the Earth any objects carrying nuclear weapons or any other kinds of weapons of mass destructio"'), the issue has been raised over and over. De facto, armed forces make an essentially immaterial use of space. They use it to collect and transport information, thanks to spacecrafts and space-based devices. In other words, space serves purposes of espionage (observation and listening) and transmission (telecommunications), both of which have had far-reaching consequences. Within the framework of their *Global Information Dominance* program, the United States have created a specialised agency, the National Imagery and Mapping Agency (NIMA), bringing together 9,000 people in charge of centralizing all the images obtained via military satellites (in particular the Key Hole and Lacrosse series, the resolution of which is said to reach 10 cm or less) and to ensure their processing. In parallel, the National Security Agency (NSA) employs 38,000 people, among whom 20,000 in the USA. As for the National Reconnaissance Office (NRO), its goal is as clear as its motto: 'Freedom's sentinel in space. One team, revolutionizing global reconnaissance.' Everyone has potentially heard of the debate around the British collaboration in the Echelon listening network within the context of the UKUSA agreement binding the United States, the United Kingdom, Canada, Australia, and New-Zealand. This network is based on tapping performances reached by combining the abilities of computers and satellites (especially the Trumpet and Vortex-2 satellites, known to have an antenna 150 meters in diameter). The Echelon network is supposedly able to spy on, sort out, decrypt, archive, and process three million satellite-transmitted telephone communications per minute, meaning phone calls, faxes, Internet messages, and all exchanges of computer data.

Up to now, the material use of space for military purposes has remained limited, either because of international agreements forbidding the deployment of certain weapons, or because of political will or technological constraints. If we make a comparison with Antartica, we note that, on that continent, all activities done with aims of national sovereignty are prohibited, as are all military weapon tests. In space, however, the 1967 treaty limits this prohibition to nuclear weapons and weapons of mass destruction. But in any case, space is used only as a passageway for intercontinental missiles without this constituting a judicial violation of its peaceful nature. In addition, the Anti-Ballistic Missile Treaty (ABM Treaty) forbids the development, testing, or deployment of sea-, air-, space-, or mobile land-based ABM systems or components.

---

### The 1967 Space Treaty and the issue of militarization

*The States Parties to this Treaty,*

*Inspired* by the great prospects opening up before humankind as a result of human's entry into outer space,

*Recognizing* the common interest of all humankind in the progress of the exploration and use of outer space for peaceful purposes [. . .]

*Desiring* to contribute to broad international cooperation in the scientific as well as the legal aspects of the exploration and use of outer space for peaceful purposes [. . .]

*Recalling* resolution 1884 (XVIII), calling upon States to refrain from placing in orbit around the Earth any objects carrying nuclear weapons or any other kinds of weapons of mass destruction or from installing such weapons on celestial bodies, which was adopted unanimously by the United Nations General Assembly on 17 October 1963,

---

*Taking account* of United Nations General Assembly resolution 110 (II) of 3 November 1947, which condemned propaganda designed or likely to provoke or encourage any threat to the peace, breach of the peace or act of aggression, and considering that the aforementioned resolution is applicable to outer space [. . .]

*Have agreed* on the following: [. . .]

## Article III

States Parties to the Treaty shall carry on activities in the exploration and use of outer space, including the Moon and other celestial bodies, in accordance with international law, including the Charter of the United Nations, in the interest of maintaining international peace and security and promoting international cooperation and understanding.

## Article IV

States Parties to the Treaty undertake not to place in orbit around the Earth any objects carrying nuclear weapons or any other kinds of weapons of mass destruction, install such weapons on celestial bodies, or station such weapons in outer space in any other manner.

The Moon and other celestial bodies shall be used by all States Parties to the Treaty exclusively for peaceful purposes. The establishment of military bases, installations and fortifications, the testing of any type of weapons, and the conduct of military maneuvers on celestial bodies shall be forbidden. The use of military personnel for scientific research or for any other peaceful purposes shall not be prohibited. The use of any equipment or facility necessary for peaceful exploration of the Moon and other celestial bodies shall also not be prohibited.

Without delving too deep into the mysteries of law and terminology, we should nonetheless be aware of the differences between the weaponization of space and its militarization. In the sense of deployment and implementation of conventional weapons, it is correct to assert that, today, space is not the stage of actual weaponization. This being said, the United States does seem to be going in that direction, in the name of its drive towards space domination and domination via space. Conversely, it must be admitted that the militarization of space has already started to a great extent, so long as this term designates the process leading to the direct contribution of space assets to operations of power projection and military operations.

If for thirty years, up to the first Gulf war, space presented an essentially strategic interest, the situation has changed with the growing importance of information in the use of modern weaponry. More than ever, mastering information is a way to guarantee the freedom of action of armed forces. Military staffs now recognise this. The conclusion is obvious: the United States has clearly militarised space to its advantage. Other countries, such as France, are following the same path to plan and conduct their operations. John Logsdon, a great connoisseur in the field of space matters and policies, is right to conclude that: 'It is high time to accept reality: the situation of the past fifty years, during which space was not only a common good but also a sanctuary sheltered from armed conflict, has reached an end.' One should even go further and question the idea, possibly too comfortable, that space is dual; namely that there is a strict separation between the civilian and military spheres.

Can one still speak of dual space when armed forced are used to relying on the services of civilian systems of remote sensing? Iraqi forces during the conflict with Iran, American forces during the Desert Storm operation, and NATO forces intervening in Bosnia and Serbia have all resorted to images from the Spot civilian satellites. Is it still possible to refer to a dual space when the GPS, originally designed for the American army, has become a spatial tool with uncountable civilian applications? Is space still dual when the French observation system Pléiades provides images to military intelligence services, ensuring them complete confidentiality for their requests as well as the safety of their communications, and above all, absolute priority over requests from civilian sectors? In short, once the barrier between civilian and military fields has been lifted, what remains of space's duality? Would it not be more appropriate to speak of twinship and to admit that similarities and collusions exist?

Acknowledging such an evolution leads us to entrust civilian space agencies with new tasks. For instance, they need to enforce the choices stated and ratified in 1967, or make officials and the public aware of the consequences for the peaceful use of space of a certain number of military practices (for example, the installation of lasers onboard satellites). They also need to make sure that data collected and transmitted by civilian satellites is not used for terrorist or unlawful purposes. The latter task is complicated by the fact that space appears here as a metaphor of capitalism: possessing technology or financial means can, in itself, guarantee access to power outside the boundaries of any juridical, political, or regulatory constraints. In this field, ethical questioning is still in its infancy.

# Bibliography

Alby, Fernand; Arnould, Jacques; Debus, André. *La pollution spatiale sous surveillance*. Paris: Ellipses, 2007.

Arnould, Jacques. *Qu'allons-nous faire dans ces étoiles? De l'éthique dans la conquête spatiale*. Paris: Bayard, 2009.

Arnould, Jacques. *La Terre d'un clic. Du bon usage des satellites*. Paris: Odile Jacob, 2010.

Arnould, Jacques. *Icarus' Second Chance. The Basis and Perspectives of Space Ethics*. New York: Springer, 2011.

Arnould, Jacques. *Le rire d'Icare. Essai sur le risque et l'aventure spatiale*. Paris: Cerf, 2013.

Arnould, Jacques. *Une perle bleue. L'espace, la Terre et le changement climatique*. Paris: Cerf, 2015.

Arnould, Jacques. *Impossible horizon. The essence of space exploration*. Adelaide: ATF Press, 2017.

Arnould, Jacques. *Oublier la Terre? La conquête spatiale 2.0*. Paris: Le Pommier, 2018.

Arnould, Jacques. *La Lune m'a dit. Cinquante ans après le premier homme sur la Lune*. Paris: Cerf, 2019.

Dator, James A. *Social Foundations of Human Space Exploration*. New York: Springer-ISU, 2012.

Hart, John. *Cosmic Commons. Spirit, Science, & Space*. Eugene (OR): Cascade Books, 2013.

McCurdy, Howard E. *Space and the American Imagination*. Washington & London: Smithsonian Institution Press, 1997.

Milligan, Tony. *Nobody Owns the Moon. The Ethics of Space Exploration*. McFarland, 2016.

O'Neill, Gerard. *The High Frontier. Human Colonies in Space*. New York: William Morrow & Company, 1977.

Schneider, Jean & Léger-Orine, Monique (éd.). *Frontières et conquête spatiale. La philosophie à l'épreuve*. Dordrecht/Boston/London: Kluwer Academic Publishers, 1988.

Schwartz, James S.J and Milligan, Tony (editors). *The Ethics of Space Exploration*. New York: Springer (Space and Society), 2016.

Wolfe, Tom. *The Right Stuff*. New York: Farrar-Straus-Giroux, 1983.

CPSIA information can be obtained
at www.ICGtesting.com
Printed in the USA
BVHW082134090222
628353BV00002B/139